PREPARING FOR THE CALCULUS

AP* EXAM

WITH
CALCULUS: Graphical, Numerical, Algebraic

FINNEY, DEMANA, WAITS, KENNEDY

WRITTEN BY
BARTON
BRUNSTING
DIEHL
HILL
TYLER
WILSON

PEARSON

Addison
Wesley

Boston   San Francisco   New York
London   Toronto   Sydney   Tokyo   Singapore   Madrid
Mexico City   Munich   Paris   Cape Town   Hong Kong   Montreal

ISBN 0-321-33574-0

10 11 12 BRR 14 13 12 11

Preparing for the AP* Calculus Exam

TABLE OF CONTENTS

About the Authors

Ray Barton teaches AP* Calculus BC at Olympus High School in Salt Lake City, Utah. He has been an Advanced Placement Calculus exam reader and is an active instructor for Teachers Teaching with Technology. He has coauthored *Advanced Placement Calculus with the TI-89* and *Differential Equations with the TI-86*. Ray received the Presidential Award for Excellence in Mathematics and Science Teaching in 1995. He is a strong proponent of using technology to teach mathematics.

John R. Brunsting taught AP* Calculus at Hinsdale Central High School in Hinsdale, Illinois. He has been an Advanced Placement Calculus exam reader and table leader, as well as an AP* Calculus Test Development Committee member. He was a consultant for the Midwest Region of the College Board. Now retired, John is a director of Illinois Advanced Placement Institutes and Mathematics & Technology Institutes, providing summer training for AP* teachers.

John J. Diehl has taught AP* Calculus and AP* Statistics at Hinsdale Central High School in Hinsdale, Illinois. He has been an Advanced Placement Statistics exam reader and table leader, as well as an AP* Statistics Test Development Committee member. John is a consultant for the Midwest Region of the College Board and a Teachers Teaching with Technology Institute instructor. John is coauthor of *Advanced Placement Calculus with the TI-89*.

Greg Hill teaches AP* Calculus BC at Hinsdale Central High School in Hinsdale, Illinois. A nearly twenty-year veteran, he has been an Advanced Placement Calculus exam reader for the last five years. As a consultant for the College Board, Greg has presented numerous AP* Calculus workshops across the Midwest. An active Teachers Teaching with Technology instructor, Greg guides teachers in the appropriate uses of technology in algebra, precalculus, and calculus.

Karyl Tyler teaches AP* Calculus AB at Hinsdale Central High School in Hinsdale, Illinois. Her twenty-eight years of teaching experience include eleven years of teaching calculus. Beyond her classroom, Karyl assists with instruction at Teachers Teaching with Technology Institutes, sharing her passion for using technology in the mathematics classroom.

Steven L. Wilson teaches AP* Calculus AB at Hinsdale Central High School in Hinsdale, Illinois. His twelve years of teaching experience include eight years of teaching calculus. A versatile mathematician, Steve shares his deep understandings with students training for rigorous mathematics oral competitions.

About Your Pearson AP* Guide

Pearson Education is the leading publisher of textbooks worldwide. With operations on every continent, we make it our business to understand the changing needs of students at every level, from kindergarten to college.

This gives us a unique insight into what kind of study materials work for students. We talk to customers every day, soliciting feedback on our books. We think that this makes us especially qualified to offer this series of AP* test prep books, tied to some of our best-selling textbooks.

We know that as you study for your AP* course, you're preparing along the way for the AP* exam. By tying the material in the book directly to AP* course goals and exam topics, we help you to focus your time most efficiently. And that's a good thing!

The AP* exam is an important milestone in your education. A high score will position you optimally for college acceptance—and possibly will give you college credits that put you a step ahead. Our goal at Pearson Education is to provide you with the tools you need to excel on the exam ... the rest is up to you.

Good Luck!

Acknowledgments

We wish to thank the late Ross Finney, and Bert Waits, Frank Demana, Dan Kennedy, and Greg Foley for precalculus and calculus textbook contributions that made linking their books to AP* Calculus objectives a nearly effortless task. Thanks also to the College Board for their long-standing commitment to educational excellence as exemplified by the AP* Calculus curriculum and assessment standards.

We are indebted to the entire Pearson Addison-Wesley publishing team for their trust, guidance, and patience with us as a writing team.

We also acknowledge the support of Dan Kennedy for his thoughtful input and wise counsel in shaping this project, Fred Gloff for working and checking all problems, and Hinsdale Township High School District 86 for its unwavering promotion of professional excellence that continues to encourage teachers to learn and to share in community.

Finally, at the most personal level, our team expresses extra special thanks to those closest to us—our families—for their encouragement, understanding, and patience during the creating, writing, rewriting, rewriting, and final rewriting phases of this project.

Part I

Introduction to the AP Calculus— AB or BC—Examination

Preparing for the AP* Calculus Exam has been written as a supplement to *Precalculus: Graphical, Numerical, Algebraic* by Demana, Waits, Foley, and Kennedy and *Calculus: Graphical, Numerical, Algebraic* by Finney, Demana, Waits, and Kennedy. The authors hope that users will find this work faithful to the themes, content, and spirit of those two texts. We also hope that users of other textbooks will recognize the broad-based approach and specific insights of this guide as useful preparation for their students.

This Book

So you are planning to take or have already enrolled in either AP* Calculus AB or AP* Calculus BC! Either AP* course will stretch and enrich your mathematics skills and knowledge and will culminate in an AP* examination.

Presently enrolled in a precalculus course?

This book will point out foundational calculus Objectives encountered in a precalculus course and put those objectives into a calculus Big Picture context, provide succinct Content explanations, give Additional Practice problems, and make Need More Help? connections to the Addison-Wesley *Precalculus* and *Calculus* textbooks.

Presently enrolled in a calculus course?

This book will point out the corresponding AP* Calculus Exam Objectives, identify the Big Picture context, provide succinct Content explanations, give Additional Practice problems and solutions, and wherever possible make Need More Help? connections to Addison-Wesley *Precalculus* and *Calculus* textbooks.

In either case, this workbook supplement to your textbook will help you clearly identify essential calculus concepts as well as improve both your understanding of such concepts and your ability to communicate your thinking so that you can be successful.

This book is not intended as a substitute for a full precalculus or calculus course or a comprehensive treatment of calculus topics. It is rather meant to act as a useful review for students who have previously studied the curriculum topics and now want to solidify their learning through recall and practice.

The Advanced Placement* Calculus AB or BC Curriculum

The AP* Calculus AB and AP* Calculus BC curricula are designed to provide courses in calculus that are equal in content to the best of collegiate courses. Each is developed by a team of educators from the high school and college communities and continues to adapt itself so that a broad base of college programs will offer credit for success in the respective examination. To find out whether your prospective institution will give credit for your AP* Calculus Examination performance, you should email or call the admissions office. You can log on to AP* Central at www.apcentral.collegeboard.com for up-to-date information about the Advanced Placement* Program.

A course in calculus is especially useful to those pursuing studies in mathematics, as well as an ever-widening variety of engineering, science, economic, and business fields.

Understanding the Advanced Placement* Calculus AB or BC Examination

AP* Calculus Examinations began in 1956 as one of the earliest examinations created by the College Board. Now offered as AP* Calculus AB and AP* Calculus BC, these examinations are given in May during the same two-week window as the other examinations. Your teacher or the AP* coordinator in your school district can give you the exact date for this year's examination. You can also check for this and other information on the AP* Central Web site. Test schedules are determined long in advance and the dates and times are not flexible. In the event of an emergency you may qualify for an alternate examination, but the qualifying conditions are extremely rigid. Plan your calendar around the examination date, and register with your teacher or coordinator to reserve an examination. Standard fees as periodically adjusted by the College Board apply.

The AP* Examinations specify calculus content for AP* Calculus AB and AP* Calculus BC under the following three headings:

- Functions, Graphs, and Limits
- Derivatives
- Integrals

Additionally, a fourth heading for AP* Calculus BC only is

- Polynomial Approximations and Series

You may bring to the examination two calculators with graphical capabilities. A complete up-to-date list of acceptable graphing calculators can be found on the College Board Web site. We highly recommend using one model throughout the calculus course and bringing that same model to the examination. The exam assumes that your calculator has built-in capabilities to

1. plot the graph of a function within an arbitrary viewing window.
2. find the zeros of functions (solve equations numerically).
3. numerically calculate the derivative of a function.
4. numerically calculate the value of a definite integral.

Many students bring a second calculator as a backup. Fresh batteries are a must! Since calculator memories need not be cleared before taking the examination, calculators may contain whatever additional programs students may desire.

The actual examination is currently formatted and has timing and grade weights as follows:

Section I. Multiple Choice Section	Total of 45 Questions/105 Minutes	50% of Test
Part A (Calculator Not Allowed)	28 questions/55 minutes	
Part B (Calculator Required)	17 questions/50 minutes	
Section II. Free Response	Total of 6 Questions/90 Minutes	50% of Test
Part A (Calculator Required)	3 questions/45 minutes	50% of the 50%
Part B (Calculator Not Allowed)	3 questions/45 minutes	50% of the 50%

The Multiple Choice and Free Response sections of the examination are equally weighted. They stand alone with a recommended break between. You should expect to work the full allotted time. You should, if at all possible, allow time in each part of each section to go back and check your work. The Multiple Choice section is done first; you should not expect to go back to it after break. Indeed, that part of the examination will be sealed and collected at the end of the 105-minute time period.

Both the Multiple Choice and Free Response sections come with separate directions for Calculator Not Allowed and Calculator Required portions. In either test section, a student cannot return to the Part A section after beginning Part B.

Calculus AB Subscore Grade for the Calculus BC Examination

Since the Calculus BC curriculum and examination encompass the Calculus AB curriculum, the Calculus BC is able to report a Calculus AB Subscore. A detailed explanation of the Calculus AB Subscore and its value can be found on the College Board Web site.

Understanding the Grading Procedure for the Advanced Placement* Calculus AB or BC Examination

The scoring of an AP* Calculus Examination is done in part by machine (Section I—Multiple Choice) and in part by calculus educators (human beings!) who read the answers as you have communicated them (Section II—Free Response). The entire test is valued at 108 points with the Multiple Choice and Free Response sections each contributing 54 points.

In the Multiple Choice section you are *awarded* 1 point for each correct answer and *penalized* one-quarter point for each incorrect answer. To convert to the possible 54-point total (reflecting 50% of the grade), the number of points is multiplied by 1.2. For example, suppose you answered 31 questions correctly but answered 10 questions incorrectly and left 4 questions blank. Here is the way your Multiple Choice section would be graded:

Number correct	31	$\times$ 1	=	31
Number incorrect	10	$\times$ -0.25	=	-2.5
Number left blank	4	$\times$ 0	=	0
POINTS EARNED			=	28.5
MULTIPLE CHOICE SCORE	= 28.5 $\times$	1.2	=	34.2

In the Free Response section, the *reader* assigns a score from 0–9 as indicated by the rubric or grading rule developed by the Chief Reader and a select team of exam leaders. A rubric is developed for each question to attain scoring consistency.

Suppose your answers to the Free Response questions are scored in the following manner:

Question 1:	8 points
Question 2:	6 points
Question 3:	8 points
Question 4:	5 points
Question 5:	6 points
Question 6:	4 points

FREE RESPONSE SCORE = 37 points

Your Total Exam score would be the sum of the Multiple Choice and Free Response scores: 34.2 + 37 = 71.2.

This composite score is then subject to the Chief Reader's interpretation of the cut points for that particular examination year's results. These cut points are set shortly after examinations are scored and are based on several factors, including statistical comparability with other years' examinations, the distributions of performance on the various parts of the current year's examination, and previous years of grade distributions.

Your results will be available from the College Board by phone, usually before mid-July, as they are mailed out to you and your school shortly after that. By August your scores are also sent to any college or university that you indicate in the general information section of your answer packet. If you do not wish to communicate your scores immediately to your institution or if you are unsure of what institution you will be attending, you need not indicate any schools; the results would then come only to your high school and you. You can have them transmitted later to your chosen schools in accordance with College Board policies. There may be a fee for this service.

Test-Taking Strategies for an Advanced Placement* Calculus Examination

You should approach the AP* Calculus Examination the same way you would any major test in your academic career. Just remember that *it is a one-shot deal*—you must be at your peak performance level on the day of the test. For that reason you should do everything that your "coach" tells you to do. In most cases your coach is your classroom teacher. It is very likely that your teacher has some experience, based on workshop information or previous students' performance, to share with you.

You should also analyze your own test-taking abilities. At this stage in your education, you probably know your strengths and weaknesses in test-taking situations. You may be very good at multiple choice but weaker in essays, or perhaps it is the other way around. Whatever your particular abilities are, evaluate them and respond accordingly. Spend more time on your weaker points. In other words, rather than spending time in your comfort zone where you need less work, try to improve your soft spots. In all cases, concentrate on *clear communication* of your strategies, techniques, and conclusions.

The following table presents some ideas in a quick and easy form. It is divided into two sections: general strategies for approaching the examination day and specific strategies for addressing particular types of questions on the examination.

General Strategies for AP* Examination Preparation

Time	DOs ☺	DON'Ts ☹
Through the Year	■ Register with your teacher/coordinator ■ Pay your fee (if applicable) on time ■ Take good notes ■ Work with others in study groups ■ Review on a regular basis ■ Evaluate your test-taking strengths and weaknesses—keep track of how successful you are when guessing	■ Procrastinate ■ Avoid homework and labs ■ Wait until the last moment to pull it together for quizzes and tests ■ Rely on others for your own progress ■ Scatter your work products (notes, labs, reviews, tests) ■ Ignore your weak areas—remediate as you go along
The Week Before	■ Combine independent and group review ■ Get tips from your teacher ■ Do lots of mixed review problems ■ Check your exam date, time, and location ■ Review the appropriate AP* Calculus syllabus (AB or BC)	■ Procrastinate ■ Think you are the only one who is stressed ■ Forget your priorities—this test is a one-shot deal
The Night Before	■ Put new batteries in your calculator; check for illegal programs ■ Lay out your clothes and supplies so that you are ready to go out the door ■ Do a short review ■ Go to bed at a reasonable hour	■ Study all night ■ Get caught without fresh batteries
Exam Day	■ Get up a little earlier than usual ■ Eat a good breakfast/lunch ■ Put some hard candy in your pocket in case you need an energy boost during the test ■ Get to your exam location 15 minutes early	■ Sleep in ■ Panic with last-minute cramming
Exam Night	■ Relax—you earned it	■ Worry—it's over

Specific Strategies to use During the AP* Examination

Question Type	DOs ☺	DON'Ts ☹
Multiple Choice	■ Underline key words and phrases; circle important information you will use ■ Look at the answer format so that you do not do unnecessary steps ■ Anticipate the likely errors for the type of question being asked—watch out for obvious choices ■ When formulas and calculations are needed, write down what you are doing so you can check your procedure ■ Eliminate as many answers as possible ■ Guess only when you are comfortable with the number of possible answers ■ Check the units of your answers	■ Rush—reading the question correctly is the key to answering the question correctly ■ Do unnecessary calculations (sometimes equations can be left in any form) ■ Fall into the traps your teacher warned you about—a question that looks too easy may have one of these traps built in ■ Scribble your work—if you need to review your procedure you wind up having to repeat it ■ Guess haphazardly ■ Spend more than 2–3 minutes on any one question—if you don't know, move on ■ Close the book until time is up
Free Response (All Questions)	■ Look over all the questions before you start and do the ones that seem easiest first ■ Read the *entire* question in *all of its parts* ■ Underline what is being asked ■ Carry out your strategy by clearly indicating your steps ■ Move on to the next part of the question if you cannot answer one part ■ Make up a reasonable answer for one part if the next part requires the previous answer ("Suppose my answer to Part A had been 20 sq. units. Using that area value and the integral $\int_0^k f(x)dx = \frac{1}{2} \cdot (20)$, I will find the number k for which the line $x = k$ divides the region equally.") ■ Write neatly, compactly, and clearly and use calculus vocabulary *correctly* ■ When useful, include graphs/sketches that illustrate your answer; be sure to label axes ■ Mark up sketches provided to illustrate your thinking ■ COMMUNICATE CLEARLY—answer the question asked and place your answer in the CONTEXT of the question ■ Review your response to make sure that it shows good *mathematical thinking*	■ Feel as though you *have* to start with #1 and finish with #5; pick the one you like best to start off ■ Forget to answer the *question asked* in your haste to write an answer ■ Begin without a plan ■ Move on until you have read what you have written for each part to make sure you have not left out important words, punctuation, or numbers ■ Get stuck on a question—if you have no idea how to proceed after thinking about it, move on ■ Scribble—a human being must be able to decipher your response ■ Waste time erasing unless space is an issue—anything crossed out will not be read as part of your answer ■ Round during computation—wait until you reach a final answer to round ■ Run on—you are likely to say something INCORRECTLY that will diminish your previously correct response ■ Assume that the size of the space provided is proportional to the answer desired

Topics from the Advanced Placement* Curriculum for Calculus AB, Calculus BC

The AP* Calculus Examination is based on the following Topic Outline. For your convenience, we have noted all Calculus AB and Calculus BC objectives with clear indications of topics required only by the Calculus BC Exam. The outline cross references each objective with our primary textbooks: *Precalculus: Graphical, Numerical, Algebraic* by Demana, Waits, Foley, and Kennedy and *Calculus: Graphical, Numerical, Algebraic* by Finney, Demana, Waits, and Kennedy.

Use this outline to keep track of your review. Be sure to cover every topic associated with the exam you are taking. Check it off when you have reviewed the topic from your text and then review the topic in this book.

Topic Outline for AP* Calculus AB and AP* Calculus BC

(Excerpted from the College Board's Course Description—Calculus: Calculus AB, Calculus BC, May 2007)

I.		Calculus Exam		Functions, Graphs, and Limits	Precalculus	Calculus
A		AB	BC	Analysis of graphs	1.2	1.2–1.6
B		AB	BC	Limits of functions (including one-sided limits)		
	B1	AB	BC	An intuitive understanding of the limiting process	10.3	2.1, 2.2
	B2	AB	BC	Calculating limits using algebra	10.3	2.1, 2.2
	B3	AB	BC	Estimating limits from graphs or tables of data	10.3	2.1, 2.2
C		AB	BC	Asymptotic and unbounded behavior		
	C1	AB	BC	Understanding asymptotes in terms of graphical behavior	1.2, 2.7	2.2
	C2	AB	BC	Describing asymptotic behavior in terms of limits involving infinity	1.2, 2.7	2.2
	C3	AB	BC	Comparing relative magnitudes of functions and their rates of change	1.2, 2.7	2.2, 2.4, 8.3
D		AB	BC	Continuity as a property of functions		
	D1	AB	BC	An intuitive understanding of continuity	1.2	2.3
	D2	AB	BC	Understanding continuity in terms of limits	1.2	2.3
	D3	AB	BC	Geometric understanding of graphs of continuous functions	2.3	2.3, 4.1–4.3
E			BC	Parametric, polar, and vector functions		10.1–10.3

II.		Calculus Exam		Derivatives	Precalculus	Calculus
A		AB	BC	Concept of the derivative		
	A1	AB	BC	Derivative presented graphically, numerically, and analytically	10.1	2.4–4.5
	A2	AB	BC	Derivative interpreted as an instantaneous rate of change	10.1	2.4
	A3	AB	BC	Derivative defined as the limit of the difference quotient	10.1	2.4–3.1
	A4	AB	BC	Relationship between differentiability and continuity	10.1	3.2
B		AB	BC	Derivative at a point		
	B1	AB	BC	Slope of a curve at a point	10.1	2.4
	B2	AB	BC	Tangent line to a curve at a point and local linear approximation	10.1	2.4, 4.5

B3	AB	BC	Instantaneous rate of change as the limit of average rate of change	10.1	2.4, 3.4
B4	AB	BC	Approximate rate of change from graphs and tables of values	10.1	2.4, 3.4
C	AB	BC	Derivative as a function		
C1	AB	BC	Corresponding characteristics of graphs of f and f'	10.1	3.1, 4.3
C2	AB	BC	Relationship between the increasing and decreasing behavior of f and the sign of f'	10.1	4.1, 4.3
C3	AB	BC	The Mean Value Theorem and its geometric consequences	2.1, 10.1	4.2
C4	AB	BC	Equations involving derivatives (Verbal descriptions are translated into equations involving derivatives and vice versa.)	—	3.4, 3.5, 4.6, 6.4, 6.5
D	AB	BC	Second derivatives		
D1	AB	BC	Corresponding characteristics of graphs of f, f' and f''	1.2	4.3
D2	AB	BC	Relationship between the concavity of f and the sign of f''	1.2	4.3
D3	AB	BC	Points of inflection as places where concavity changes	—	4.3
E	AB	BC	Applications of derivatives		
E1	AB	BC	Analysis of curves, including the notions of monotonicity and concavity	2.3	4.1–4.3
E2		BC	Analysis of planar curves given in parametric form, polar form, and vector form, including velocity and acceleration vectors	—	10.1–10.3
E3	AB	BC	Optimization, both absolute (global) and relative (local) extrema	1.6	4.3, 4.4
E4	AB	BC	Modeling rates of change, including related rate problems	—	4.6
E5	AB	BC	Use of implicit differentiation to find the derivative of an inverse function	—	3.7
E6	AB	BC	Interpretation of the derivative as a rate of change in varied applied contexts, including velocity, speed, and acceleration	—	3.4
E7	AB	BC	Geometric interpretation of differential equations via slope fields and the relationship between slope fields and solution curves for differential equations	—	6.1
E8		BC	Numerical solution, of differential equations using Euler's method	—	6.1
E9		BC	L'Hôpital's Rule, including its use in determining limits and convergence of improper integrals and series	—	8.2, 9.5
F	AB	BC	Computation of derivatives		
F1	AB	BC	Knowledge of derivatives of basic functions, including power, exponential, logarithmic, trigonometric, and inverse trigonometric functions	—	3.3, 3.5
F2	AB	BC	Basic rules for the derivative of sums, products, and quotients of functions		3.3, 3.8, 3.9
F3	AB	BC	Chain rule and implicit differentiation	1.4	3.6, 3.7
F4		BC	Derivatives of parametric, polar, and vector functions	—	10.1–10.3

III.	Calculus Exam		Integrals	Precalculus	Calculus
A			Interpretations and properties of definite integrals		
A1	AB	BC	Definite integral as a limit of Riemann sums over equal subdivisions	–	5.1, 5.2
A2	AB	BC	Definite integral of the rate of change of a quantity over an interval interpreted as the change of the quantity over the (interval: $\int_a^b f'(x)\,dx = f(b) - f(a)$	–	5.1, 5.4
A3	AB	BC	Basic properties of definite integrals (Examples include additivity and linearity.)	–	5.2, 5.3
B			Applications of integrals		
B1a	AB	BC	Appropriate integrals are used in a variety of applications to model physical, biological, or economic situations. Although only a sampling of applications can be included in any specific course, students should be able to adapt their knowledge and techniques.	–	5.4, 5.5, 6.4, 6.5, 7.1–7.5
B1b		BC	Appropriate integrals are used ... specific applications should include ... finding the area of a region (including a region bounded by polar curves) ... the distance traveled by a particle along a line, and the length of a curve (including a curve given in parametric form).	–	7.4, 10.1, 10.3
C			Fundamental Theorem of Calculus		
C1	AB	BC	Use of the Fundamental Theorem to evaluate definite integrals	–	5.4
C2	AB	BC	Use of the Fundamental Theorem to represent a particular antiderivative, and the analytical and graphical analysis of functions so derived	–	5.4, 6.1
D			Techniques of antidifferentiation		
D1	AB	BC	Antiderivatives following directly from derivatives of basic functions	–	4.2, 6.1, 6.2
D2a	AB	BC	Antiderivatives by **substitution of variables** (including change of limits for definite integrals)	–	6.2
D2b		BC	Antiderivatives by **parts,** and simple partial fractions (nonrepeating linear factors only)	3.3, 3.4, 7.4	6.3, 6.5
D3		BC	Improper integrals (as limits of definite integrals)	–	8.3
E			Applications of antidifferentiation		
E1	AB	BC	Finding specific antiderivatives using initial conditions, including applications to motion along a line	–	6.1, 7.1
E2	AB	BC	Solving separable differential equations and using them in modeling. (In particular, studying the equations $y' = ky$ and exponential growth)	–	6.4
E3		BC	Solving logistic differential equations and using them in modeling	–	6.5
F			Numerical approximations to definite integrals		
F1	AB	BC	Use of Riemann and trapezoidal sums to approximate definite integrals of functions represented algebraically, graphically, and by tables of values	–	5.2, 5.5

IV.	Calculus Exam	Polynomial Approximations and Series	Precalculus	Calculus
A		Concept of series		
A1	BC	A series is defined as a sequence of partial sums, and convergence is defined in terms of the limit of the sequence of partial sums. Technology can be used to explore convergence or divergence.	—	9.1
B		Series of constants		
B1	BC	Motivating examples, including decimal expansion	—	9.1
B2	BC	Geometric series with applications	—	9.1
B3	BC	The harmonic series	—	9.5
B4	BC	Alternating series with error bound	—	9.5
B5	BC	Terms of series as areas of rectangles and their relationship to improper integrals, including the integral test and its use in testing the convergence of p-series	—	9.5
B6	BC	The ratio test for convergence or divergence	—	9.4
B7	BC	Comparing series to test for convergence and divergence	—	9.4
C		Taylor series		
C1	BC	Taylor polynomial approximation with graphical demonstration of convergence (For example, viewing graphs of various Taylor polynomials of the sine function approximating the sine curve.)	—	9.2
C2	BC	Maclaurin series and the general Taylor series centered at $x = a$	—	9.2
C3	BC	Maclaurin series for the functions e^x, $\sin x$, $\cos x$, and $1/(1 - x)$	—	9.2
C4	BC	Formal manipulation of Taylor series and shortcuts to computing Taylor series including substitution, differentiation, antidifferentiation, and the formation of new series from known series	—	9.1, 9.2
C5	BC	Functions defined by power series	—	9.1, 9.2
C6	BC	Radius and interval of convergence of power series	—	9.1, 9.4, 9.5
C7	BC	Lagrange error bound for Taylor polynomials	—	9.3

Part II

Precalculus Review of Calculus Prerequisites

If you are presently in a precalculus course, perhaps you have felt somewhat overwhelmed—so much to learn, so little time! All teachers know that many students struggle distinguishing important course content from the extremely important! *Preparing for the AP* Calculus Exam* will help you address the essential question:

What are the most important things to pay attention to in a precalculus course in order to be well prepared for an AP* Calculus course?

Precalculus—A Preparation for Calculus!

As the word implies, precalculus is a preparation for calculus. Whether your precalculus course uses *Precalculus: Graphical, Numerical, Algebraic* by Demana, Waits, Foley, and Kennedy (hereafter *Precalculus*) or some other textbook, the course should provide a good foundation for advanced mathematical study. If an AP* Calculus course is in your future, you should know the specific content, concepts, and skills taught in a precalculus course that will be encountered frequently throughout your calculus course.

The College Board AP* Calculus Course Description booklet, May 2007, describes the following prerequisites needed for calculus:

> Before studying calculus, all students should complete four years of secondary mathematics designed for college-bound students: courses in which they study algebra, geometry, trigonometry, analytic geometry, and elementary functions. These functions include those that are linear, polynomial, rational, exponential, logarithmic, trigonometric, inverse trigonometric, and piecewise defined. In particular, before studying calculus, students must be familiar with the properties of functions, the algebra of functions, and the graphs of functions. Students must also understand the language of functions (domain and range, odd and even, periodic, symmetry, zeros, intercepts, and so on) and know the values of the trigonometric functions of the numbers 0, $\frac{\pi}{6}, \frac{\pi}{4}, \frac{\pi}{3}, \frac{\pi}{2}$, and their multiples.

[AP Calculus Course Description, May 2007]

This precalculus section is structured to identify the important precalculus Objectives, describe the importance of each in a Big Picture calculus context, provide succinct Content explanations, give Additional Practice problems, and point to resources if you Need More Help. Ten topics—identified as Calculus Prerequisite Knowledge—are listed below and cross referenced to the *Precalculus* textbook.

AP* Preparation Topic	Calculus Prerequisite Knowledge	Precalculus Textbook
0	Basic functions	1.3
1	Functions	1.3
2	Transformations	1.5
3	Polynomial functions	2.3
4	Rational Functions	2.7
5	Exponential functions	3.1, 3.2
6	Sinusoidal functions	4.4
7	Other trigonometric functions	4.5
8	Inverse trigonometric functions	4.7
9	Parametric relations	6.3
10	Numerical derivatives and integrals	10.4

For those using *Precalculus,* there are excellent features and calculus cues with the book. A few of them are noted below:

■ Chapter P—Prerequisites. This wonderfully concise opening chapter identifies mathematical content, algebraic manipulation skills, and technology-related knowledge needed in both precalculus and calculus courses.

■ Chapter 1, Section 1 provides a problem-solving process that incorporates the traditional algebraic methods as well as the graphical and numerical methods associated with graphing utilities.

■ Chapter 1, Section 3 highlights the twelve basic functions that are used throughout calculus and captures their respective distinctives. This display is so useful that it is reprinted on the next pages of this book. Knowledge of these functions will be incredibly important because they are used constantly to illustrate calculus concepts and to model real-world phenomenon (e.g., linearity, exponential growth, or periodicity).

■ Throughout *Precalculus,* many examples and topics are marked with an icon, 📖 , to point out concepts that foreshadow calculus concepts such as limits, extrema, asymptotes, and continuity. For your convenience, the table that follows shows each *Precalculus* icon location and references the AP* Calculus Topic Outline found in Part I.

Precalculus Icon Location	*Precalculus* Icon Reference Description	AP* Calculus Objective Outline Code
1.1, p. 74	Solving equations algebraically	I. A
1.2, p. 91	Continuity	I. D
1.2, p. 92	Increasing/decreasing functions	II. C2
1.2, p. 100	End behavior	I. C
1.3, p. 110	Analyzing functions graphically	I. A
1.4, p. 121	Decomposing functions	II. F3
2.1, p. 172	Rate of change	II. A & B
2.1, p. 179	Maximum revenue	II. E3
2.1, pp. 180–181	Free fall	II. E6
2.3, p. 202	Local extrema	II. E3
2.3, p. 206	Intermediate value	I. D1
2.7, p. 248	Rational functions	I. C1
2.7, pp. 251–252	Optimization applications	II. E3
3.1, p. 277	Exponential change	II. E4
3.1, p. 281	Exponential base e	II. E4
3.3, p. 303	Base e logarithms	II. E4
3.4, pp. 312–313	Change of base; base b	II. E4
3.5, p. 321	Exponential equations	II. E4
4.1, p. 354	Angular/linear motion	II. E6
4.7, p. 419	Inverse trig function composition	II. E4
5.1, p. 446	Pythagorean identities	II. F1
5.2, p. 459	Identities in calculus	II. F1
5.4, p. 472	Power-reducing identities	II. F1
6.3, p. 528	Motion of objects	II. E6
7.4, p. 608	Partial fraction decomposition	I. C2
8.6, p. 690	Quadric surfaces	II. F3
9.4, p. 747	Infinite series	I. B

Twelve Basic Functions

The Identity Function

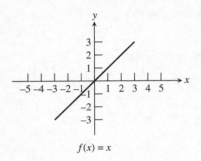

$$f(x) = x$$

Interesting fact: This is the only function that acts on every real number by leaving it alone.

The Squaring Function

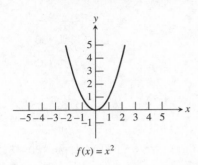

$$f(x) = x^2$$

Interesting fact: The graph of this function, called a parabola, has a reflection property that is useful in making flashlights and satellite dishes.

The Cubing Function

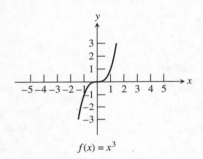

$$f(x) = x^3$$

Interesting fact: The origin is called a "point of inflection" for this curve because the graph changes curvature at that point.

The Reciprocal Function

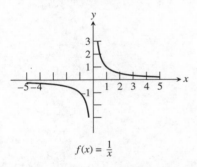

$$f(x) = \frac{1}{x}$$

Interesting fact: This curve, called a hyperbola, also has a reflection property that is useful in satellite dishes.

The Square Root Function

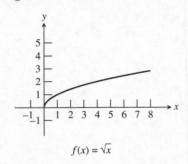

$$f(x) = \sqrt{x}$$

Interesting fact: Put any positive number into your calculator. Take the square root. Then take the square root again. Then take the square root again, and so on. Eventually you will always get 1.

The Exponential Function

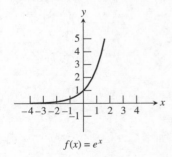

$$f(x) = e^x$$

Interesting fact: The number e is an irrational number (like π) that shows up in a variety of applications. The symbols e and π were both brought into popular use by the great Swiss mathematician Leonhard Euler (1707–1783).

The Natural Logarithm Function

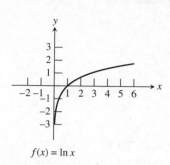

$$f(x) = \ln x$$

Interesting fact: This function increases very slowly. If the x-axis and y-axis were both scaled with unit lengths of one inch, you would have to travel more than two and a half miles along the curve just to get a foot above the x-axis.

The Sine Function

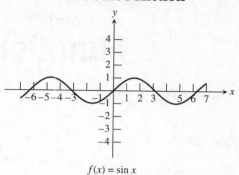

$$f(x) = \sin x$$

Interesting fact: This function and the sinus cavities in your head derive their names from a common root: the Latin word for "bay." This is due to a 12th-century mistake made by Robert of Chester, who translated a word incorrectly from an Arabic manuscript.

The Cosine Function

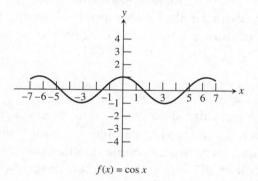

$$f(x) = \cos x$$

Interesting fact: The local extrema of the cosine function occur exactly at the zeros of the sine function, and vice versa.

The Absolute Value Function

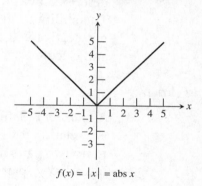

$$f(x) = |x| = \text{abs } x$$

Interesting fact: This function has an abrupt change of direction (a "corner") at the origin, while the other functions are all "smooth" on their domains.

The Greatest Integer Function

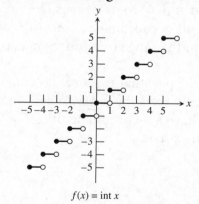

$$f(x) = \text{int } x$$

Interesting fact: This function has a jump discontinuity at every integer value of x. Similar-looking functions are called *step functions*.

The Logistic Function

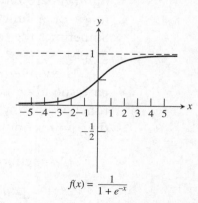

$$f(x) = \frac{1}{1 + e^{-x}}$$

Interesting fact: There are two horizontal asymptotes, the x-axis and the line $y = 1$. This function provides a model for many applications in biology and business.

Functions

Objectives:

• Identify the domain and range for a given function.
• Compute the *x*-intercepts and the *y*-intercept for a given function.

Big Picture

Functions are the key mathematical concept in precalculus and calculus. You should understand the definition of a function and how functions are described by equations and tables. Every function has a corresponding graph. The ability to look at functions from an algebraic, numerical, and graphical perspective will be a great aid in understanding the concepts of precalculus and calculus.

Content and Practice

A *relation* is defined a set of ordered pairs, usually of real numbers. A *function* is a relation for which each ordered pair (x, y) has a unique *x*-coordinate; that is, no two pairs may have the same *x*-coordinate. Some functions are not written as (x, y) pairs, but the definition still holds: the first coordinate, whatever the variable, must be unique.

The *domain* of a function is the set of all *x*-coordinates (first coordinates). It is understood that if a domain is not given for a function, we should select the largest domain possible—that is, all possible real values of *x* that can be used. Two very common reasons that restrict domains to only certain real numbers are zero denominators and square roots of negative numbers. The *range* of a function is the set of all *y*-coordinates (second coordinates). Sometimes the easiest way to confirm the range of a function is to inspect its graph.

The *y-intercept* of a function is the pair that has an *x*-coordinate of 0; the corresponding point on the graph intersects the *y*-axis. The *x-intercepts* of a function are any ordered pairs that have a *y*-coordinate of 0. The corresponding points on the graph intersect the *x*-axis.

1. A function *f* is defined as $f(x) = \sqrt{x + 4}$.

   (A) Identify the domain of the function.

(B) Identify the range of the function.

(C) Compute the y-intercept.

(D) Compute the x-intercept(s).

(E) Sketch a graph of the function.

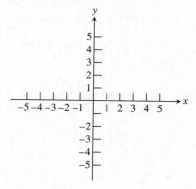

2. A function f is defined as $f(x) = x^2 - 2x - 3$.

(A) Identify the domain of the function.

(B) Identify the range of the function.

(C) Compute the y-intercept.

(D) Compute the x-intercept(s).

(E) Sketch a graph of the function.

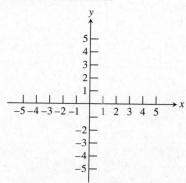

1. The domain for the function $f(x) = \dfrac{1}{\sqrt{x-2}}$ is

   (A) $x \geq 0$ (B) $x < 2$ (C) $x \leq 2$

   (D) $x > 2$ (E) $x \geq 2$

2. For each function below:

   (A) Sketch a graph. (Can you do it without the use of a calculator?)

   (B) Identify the domain.

   (C) Identify the range.

   i. $f(x) = \frac{1}{x}$ 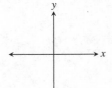 Domain:
   Range:

   ii. $g(x) = \sqrt{x}$ 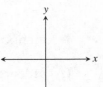 Domain:
   Range:

   iii. $h(x) = \dfrac{1}{x^2 - 9}$ 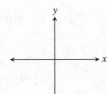 Domain:
   Range:

   iv. $k(x) = \dfrac{1}{\sqrt{x}}$ 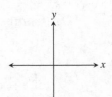 Domain:
   Range:

   v. $p(x) = x^2$ 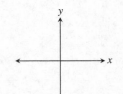 Domain:
   Range:

vi. $q(x) = \sin x$

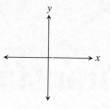

Domain:

Range:

vii. $s(x) = \tan x$

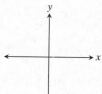

Domain:

Range:

3. Determine the ordered pairs of all intercepts of $f(x) = x^3 - 9x$.

4. The range of the piecewise function defined by

$$f(x) = \begin{cases} (x-1)^2, & x < 2 \\ 2x - 3, & x > 2 \end{cases} \text{ is}$$

(A) {all real numbers} (B) $\{y > 1\}$ (C) $\{y < 1\}$
(D) $\{y \neq 1\}$ (E) $\{y \geq 0\}$

Need More Help With . . .

Understanding functions?

See . . .

Precalculus, Section 1.3

Transformations

Objectives:

- Write the function rule, given a parent function and a set of transformations.
- Identify the transformations for a given function rule and parent function.

Big Picture

In addition to being familiar with a basic set of parent functions and their corresponding graphs, you should be able to describe important features of other functions in the family of functions associated with a particular parent function.

Content and Practice

We relate graphs using **transformations,** which are functions that map real numbers to real numbers. By acting on the x-coordinates and y-coordinates of points, transformations change graphs in predictable ways.

The key transformations are translations, reflections, stretches and shrinks.

Translations

Let c be a positive real number. Then the following transformations result in translations (shifts) of the graph of $y = f(x)$:

Horizontal translations

$y = f(x - c)$ a translation to the right by c units

$y = f(x + c)$ a translation to the left by c units

Vertical translations

$y = f(x) + c$ a translation upward of c units

$y = f(x) - c$ a translation downward of c units

Reflections

The following transformations result in reflections of the graph of $y = f(x)$:

Across the x-axis

$$y = -f(x)$$

Across the y-axis

$$y = f(-x)$$

Stretches and Shrinks

Let c be a positive real number. Then the following transformation result in stretches or shrinks of the graph of $y = f(x)$:

Horizontal stretches or shrinks

$$y = f\left(\frac{x}{c}\right)$$ a stretch by a factor of c, if $c > 1$
     a shrink by a factor of c, if $c < 1$

Vertical stretches or shrinks

$$y = c \cdot f(x)$$ a stretch by a factor of c, if $c > 1$
     a shrink by a factor of c, if $c < 1$

When two vertical transformations are used, the order of operations prevails. For example, a vertical stretch would be done before a vertical shift. However, when two horizontal transformations are used, it is generally easier to describe the transformation in reverse—for example, a shift before a stretch.

1.   A function g is defined as $g(x) = 2\sqrt{x + 4} + 3$.

    (A)   Identify the domain of the function.

    (B)   Identify the range of the function.

    (C)   Describe the transformations that show how the graph of this function is obtained from the graph of the parent function $f(x) = \sqrt{x}$.

(D) Sketch a graph of the function.

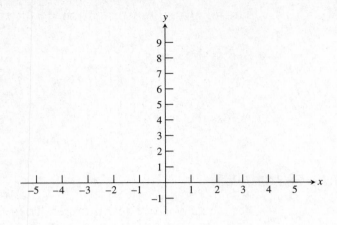

2.   The graph of a function $g(x)$ is obtained from the graph of the parent function $f(x) = x^2$ by an x-axis reflection, a vertical stretch by 3, a vertical shift down 4, and a horizontal shift right 1.

(A)   Write the equation that describes the rule for the function.

(B)   Sketch a graph of the function.

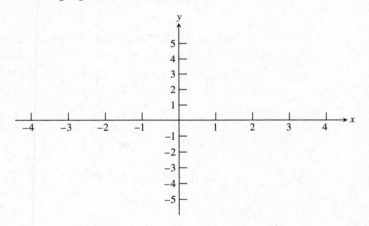

(C)   Identify the range of the function.

3. The equation of the graph of $y = \sin x$ reflected in the x-axis is

(A) $y = \sin(-x)$

(B) $y = -\sin x$

(C) $y = \sin(x - 1)$

(D) $y = -\sin(-x)$

(E) $x = \sin y$

Additional Practice

1. The graph of $f(x) = |x|$ is shown in the figure.

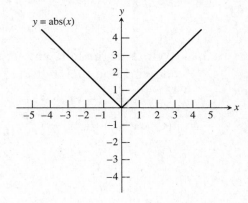

Sketch a graph of each function.

(A) $y = -f(x - 1) + 2$

(B) $y = 2f\left(\dfrac{x}{3} + 1\right)$

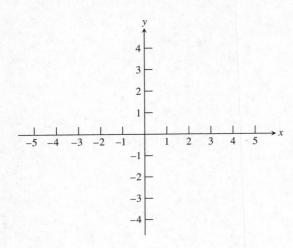 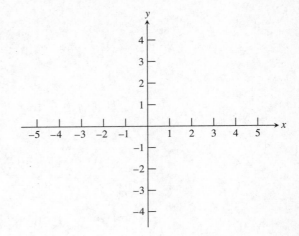

2. Which of the following represents a vertical shift up 3 and a horizontal shift left 4 of $f(x) = |x|$?

(A) $g(x) = |x - 4| - 3$ (B) $g(x) = |x + 4| - 3$

(C) $g(x) = |x - 4| + 3$ (D) $g(x) = |x + 4| + 3$

(E) $g(x) = |x + 3| - 4$

3. Which of the following represents the graph of $y = f(x)$ first shifted down 1 and then reflected in the x-axis?

(A) $y = f(-x) - 1$ (B) $y = -(f(x) - 1)$

(C) $y = -f(x) - 1$ (D) $y = |f(x - 1)|$

(E) $y = f(-x) + 1$

Need More Help With . . .

Transformations?

See . . .

Precalculus, Section 1.5

Polynomial Functions

 Objectives:

- Determine the possible number of real zeros for a polynomial function.
- Determine the possible number of extrema for a polynomial function.
- Determine the end behavior of a polynomial function.
- Determine the intervals where a polynomial function is increasing or decreasing.
- Compute zeros and extrema of a polynomial function.

Big Picture

You should be familiar with the family of functions known as *polynomial functions*. These functions are continuous. The degree and the leading coefficient describe specific patterns of increasing/decreasing and end behavior as well as the possible number of zeros and extreme points.

Content and Practice

For this section we are assuming polynomial functions with real coefficients and unrestricted domains.

A polynomial function of odd degree always has at least one real zero. (Since function "zeros" equate to graphical x-intercepts, that also guarantees that the graph of a polynomial of odd degree must intersect the x-axis at least once.) Because nonreal zeros occur in conjugate pairs, the possible number of real zeros increases by two, up to the degree of the function. Thus, a third-degree polynomial function has one or three real zeros, and a fifth-degree polynomial function has one, three, or five real zeros. Similarly, a fourth-degree polynomial function has zero, two, or four real zeros.

1. Use a graphing calculator to graph a variety of cubic equations:
   $y = x^3, y = x^3 - 3, y = x^3 - x^2, y = x^3 - x^2 - 2$, and others of your choice.

   (A) Note the number of zeros of each function.

   (B) Note the number of x-intercepts of the corresponding graph.

(C) Note the number of extrema of each function.

(D) Note the end behaviors of each function.

2. Use a graphing calculator to graph a variety of fourth-degree equations: $y = x^4, y = x^4 - x, y = x^4 - x^2, y = x^4 - x^3 + 2x^2 - x - 1$, and others of your choice.

(A) Note the number of zeros of each function.

(B) Note the number of x-intercepts of the corresponding graph.

(C) Note the number of extrema of each function.

(D) Note the end behaviors of each function.

When a function such as $y = (x - 5)^3$ or $y = x^5$ has a single factor that is repeated, in general form $(x - c)^m$, we say that the respective zero has multiplicity m. So for $y = (x - 5)^3$ in which $(x - 5)$ is a repeated factor, 5 is a zero of multiplicity 3. For $y = x^5$, 0 is a zero of multiplicity 5.

The end behavior of a polynomial function of even degree is the same as x approaches both positive and negative infinity. When the leading coefficient is positive, the end behavior is: as $x \to \infty, y \to \infty$, and as $x \to -\infty, y \to \infty$. When the leading coefficient is negative, the end behavior is: as $x \to \infty, y \to -\infty$, and as $x \to -\infty, y \to -\infty$. Because of this behavior, a polynomial function of even degree has an odd number of extrema. For example, a fourth-degree function has three or one extrema.

The end behavior of a polynomial function of odd degree is different as x approaches both positive and negative infinity. When the leading coefficient is positive, the end behavior is: as $x \to \infty, y \to \infty$, and as $x \to -\infty, y \to -\infty$. When the leading coefficient is negative, the end behavior is: as $x \to \infty, y \to -\infty$, and as $x \to -\infty, y \to \infty$. Because of this behavior, a polynomial function of odd degree has an even number of extrema. For example, a fifth-degree function has four, two, or no extrema.

3. A function f is defined as $f(x) = 2x^3 - 9x^2 - 24x + 31$.

(A) Identify the *possible* number of real zeros.

(B) Identify the *possible* number of extreme points.

(C) Predict the end behavior of f.

(D) Compute all real zeros.

(E) Compute the coordinates of all extrema.

(F) Describe the rising/falling behavior of the graph.

(G) Sketch the graph of f. Confirm with a graphing calculator.

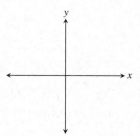

4. If $f(x) = (3x + 1)(x - 1)^3(x^2 + 4)$, then $f(x)$ has how many distinct real zeros?

(A) 1   (B) 2   (C) 3   (D) 4   (E) 6

1. A function f is defined as $f(x) = x^4 - 8x^2 + 7$.

   (A) Identify the possible number of real zeros.

   (B) Identify the possible number of extreme points.

   (C) Predict the end behavior of f.

   (D) Compute all real zeros.

   (E) Compute the coordinates of all extrema.

   (F) Describe the rising/falling behavior of the graph.

   (G) Sketch the graph of f. Confirm with a graphing calculator.

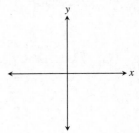

 2. Which of the following describes the possible number of real zeros and extrema for a fifth-degree polynomial function?

   (A) 5 real zeros; 4 extrema

   (B) 4, 2, or 0 real zeros; 3 or 1 extrema

   (C) 5 real zeros; 4, 2, or 0 extrema

   (D) 5, 3, or 1 real zeros; 4 extrema

   (E) 5, 3, or 1 real zeros; 4, 2, or 0 extrema

Need More Help With . . .	*See . . .*
Polynomial functions?	*Precalculus,* Section 2.3

Rational Functions

 Objectives:

- Determine the domain for a rational function.
- Determine the vertical asymptotes and removable discontinuities.
- Determine the end behavior for a rational function, including horizontal asymptotes.

Big Picture

You should be familiar with the family of functions known as *rational functions*. These functions are usually continuous over their domain, but have values that are not in the domain. Thus, they are often not continuous everywhere. You should be able to determine the behavior of the function at these points. In addition, you should be able to describe the end behavior. All of this information should be used to sketch a graph. The ability to look at these functions from an algebraic, numerical, and graphical perspective will be a great aid in understanding the calculus concept of limits.

Content and Practice

A rational function is a ratio of two polynomial functions.

$$f(x) = \frac{5x}{x + 3} \qquad g(x) = \frac{x^2 - 4}{x - 2} \qquad h(x) = \frac{x^2 - 3x - 18}{x^2 - 6x} \qquad k(x) = \frac{x + 1}{(x + 1)^3}$$

▋ *Domains:* If the function in the denominator has any real zeros, these values are not in the domain.

- The domain of $f(x)$ does not include -3.
- The domain of $g(x)$ does not include 2.
- The domain of $h(x)$ does not include 0 or 6.
- The domain of $k(x)$ does not include -1.

▋ *Removable discontinuities:* You should determine if the function has a removable discontinuity. You could factor the numerator and denominator separately. If a linear factor in the denominator appears at least as often in the numerator, then the function will have a removable discontinuity at the value that makes that factor zero.

- $g(x)$ has a removable discontinuity at 2 because it has a common factor of $(x - 2)$ in its numerator and denominator. Since $\lim_{x \to 2} g(x) = 4$, (2,4) is identified as the ordered pair associated with the removable discontinuity.

- $h(x)$ has a removable discontinuity at 6 for a similar reason. Since $\lim_{x \to 6} g(x) = \frac{3}{2}$, $\left(6, \frac{3}{2}\right)$ is identified as the ordered pair associated with the removable discontinuity.

▮ *Vertical asymptotes:* You should determine if the function has a vertical asymptote. Again, considering the linear factors in the numerator and denominator, if the factor is in the denominator only, or is a factor of the denominator more times than in the numerator, then the function has a vertical asymptote at the value that makes the factor zero.

- $f(x)$ has a vertical asymptote at $x = -3$.
- $h(x)$ has a vertical asymptote at $x = 0$.
- $k(x)$ has a vertical asymptote at $x = -1$.

In the language of calculus, we wish to determine the values that are not in the domain. Then, we investigate the limit as x approaches that value. If the limit is finite, then we have a removable discontinuity; if it is infinite, then there is an asymptote.

▮ *End behavior:* You should also investigate the end behavior. If the degree of the numerator is less than the degree of the denominator, then the end behavior is zero and the function has a horizontal asymptote, $y = 0$.

- $k(x)$ has a horizontal asymptote, $y = 0$.

▮ *Horizontal asymptotes:* If the degree of the numerator is the same as the degree of the denominator, the function has a horizontal asymptote but not at zero. The value is determined by the ratio of the leading coefficients.

- $f(x)$ has a horizontal asymptote, $y = 5$.
- $h(x)$ has a horizontal asymptote, $y = 1$.

▮ *Slant asymptotes:* If the degree of the numerator is one higher than the degree of the denominator, the function has a linear asymptote but it is not horizontal; rather, it is a slant (or oblique) asymptote.

- **Parabolic asymptotes:** If the degree of the numerator is more than one higher than the degree of the denominator, the function has an end behavior asymptote but it is not linear. For example, if the degree of the numerator is two higher, expect a parabolic asymptote.

In the case of either slant and parabolic asymptotes, the end behavior asymptote is determined by the quotient (without remainder) of the ratio of the two polynomials.

- $g(x)$ has a slant asymptote, $y = x + 2$.
- None of the functions f, g, h, or k, has a parabolic asymptote.

1. A function f is defined as $f(x) = \dfrac{x^2 + 3x - 18}{x^2 - 9}$.

   (A) Write the domain.

   (B) Write the ordered pair for any removable discontinuities.

   (C) Write the equation for any vertical asymptotes.

   (D) Write the equation for any horizontal asymptotes.

   (E) Describe the end behavior of f.

   (F) Compute the y-intercept.

(G) Sketch the graph of f.

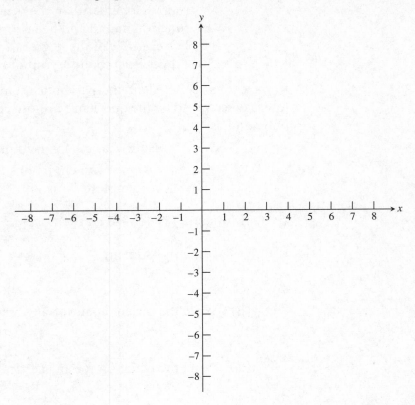

2. $f(x)$ is defined as $f(x) = \dfrac{x^2}{x - 1}$.

(A) Complete the table of values for $f(x)$.

x	$f(x)$
0	
0.9	
0.99	
0.999	
1.001	
1.01	
1.1	
2	

(B) Referring to the table of values, describe the graph of f near $x = 1$.

1. A function f is defined as $f(x) = \dfrac{x^2 + x + 3}{x - 1}$.

   (A) Write the domain.

   (B) Write the ordered pair for any removable discontinuities.

   (C) Write the equation for any vertical asymptotes.

   (D) Write the equation for any other asymptotes.

   (E) Describe the end behavior of f.

   (F) Compute the y-intercept.

   (G) Sketch the graph of f. Confirm with a graphing calculator.

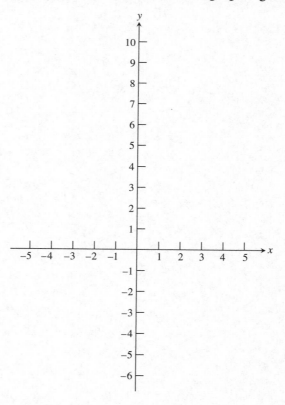

2. Which of the following best describes the behavior of the function $f(x) = \dfrac{x^2 - 2x}{x^2 - 4}$ at the values not in its domain?

(A) One vertical asymptote, no removable discontinuities

(B) Two vertical asymptotes

(C) Two removable discontinuities

(D) One removable discontinuity, one vertical asymptote, $x = 2$

(E) One removable discontinuity, one vertical asymptote, $x = -2$

Need More Help With . . .

Rational functions?

See . . .

Precalculus, Section 2.7

Exponential Functions

 Objectives:

• Determine the range for an exponential function.
• Determine the end behavior for an exponential function, including
  horizontal asymptotes.

Big Picture

You should be familiar with the family of functions known as *exponential
functions.* These functions usually have a domain of all real numbers and are
continuous over their domain. You should be able to describe the end behavior, which often includes a horizontal asymptote in one direction. All of this
information should be used to sketch a graph.

Content and Practice

A basic exponential function has an equation of the form $f(x) = a \cdot b^x$.
Transformed functions may have additional coefficients, such as
$f(x) = a \cdot b^{x-h} + k$. For $a = \frac{1}{2}$, $b = 3$, $h = 0$, and $k = -4$, the function is
$f(x) = \frac{1}{2} 3^x - 4$ and its graph is shown in the figure.

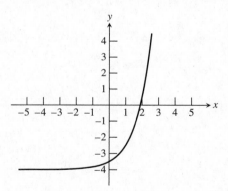

The end behavior of such functions will always be infinite as $x \rightarrow \infty$ (if
$b > 1$) or $x \rightarrow -\infty$ (if $0 < b < 1$) and have a horizontal asymptote for the
other direction. Related behavior will mean that the function is either increasing or decreasing and that the graph of the function is rising or falling.

Exponential functions model many real phenomena, such as population growth, half-life decay, and Newton's law of cooling. For many of these models the y-intercept represents an initial value (at time 0) for the variable that y represents.

Example: A town's population is 3000 and projected to double every 10 years. An exponential model for the population, P, expected in d decades, would be $P = 3000 \cdot 2^d$.

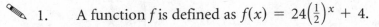 1. A function f is defined as $f(x) = 24\left(\frac{1}{2}\right)^x + 4$.

(A) Compute the y-intercept.

(B) Compute $f(x)$ for $x \in \{-2, -1, 1, 2\}$.

(C) Is f increasing or decreasing? Explain.

(D) Write the equation for the horizontal asymptote.

(E) Describe the end behavior of f.

(F) Sketch the graph of f.

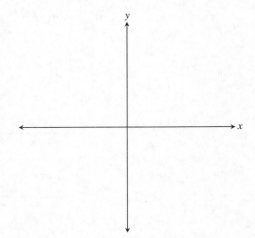

Additional Practice

1. A function f is defined as $f(x) = 3 \cdot 2^x$.

    (A) Compute the y-intercept.

    (B) Compute $f(x)$ for $x \in \{-2, -1, 1, 2\}$.

    (C) Is f increasing or decreasing?.

    (D) Write the equation for the horizontal asymptote.

    (E) Describe the end behavior of f.

    (F) Sketch the graph of f.

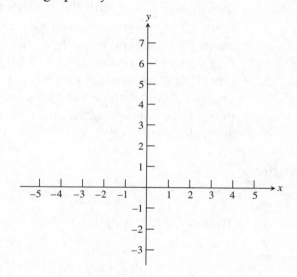

2. A radioactive substance decays so that half of the substance decays every 2 minutes. If 100 g of the substance are present initially, how many grams will be present after 4 minutes and 8 minutes, respectively?

    (A) 6.25, 0.390625

    (B) 25, 6.25

    (C) 25, 12.5

    (D) 50, 12.5

    (E) 50, 25

Need More Help With . . .

Exponential functions?

See . . .

Precalculus, Sections 3.1–3.3

Sinusoidal Functions

 Objectives:

- Determine the equations for a sinusoidal function for a particular graph.
- Determine the transformations for a sinusoidal function, given an equation.

Big Picture

Functions are the key mathematical concept in precalculus and calculus. You should be familiar with the family of functions known as *trigonometric functions*. The first type we investigate are *sinusoidal functions*, defined as a *sine* or *cosine function*. You should be able to sketch a graph for these types of functions, given the equation. You should also be able to describe the transformations that show how the graph compares to the basic sine or cosine graph. The ability to look at these functions from an algebraic, numerical, and graphical perspective will be a great aid in understanding the concepts of precalculus and calculus.

Content and Practice

The two basic sinusoidal functions are $y = \sin(x)$ and $y = \cos(x)$. Transformed functions will have additional coefficients, such as $y = a \cdot \sin[b(x - h)] + k$. The coefficient a determines the amplitude, the same transformation as a vertical stretch. The coefficient k determines the vertical shift. The coefficient h determines the horizontal shift, which in trigonometry is often referred to as a phase shift. The coefficient b determines the horizontal stretch or shrink. Because sinusoidal graphs are periodic, this coefficient also determines the period of the graph, found by computing $\frac{2\pi}{|b|}$.

1. A function f is defined as $f(x) = 3 \cos\left[\frac{1}{2}\left(x - \frac{\pi}{4}\right)\right] + 1$.

    (A) Determine the amplitude.

    (B) Determine the vertical shift.

    (C) Determine the range of the function.

(D) Determine the horizontal shift.

(E) Determine the period.

(F) Write the coordinates of two local maximum and two local minimum points.

(G) Sketch the graph.

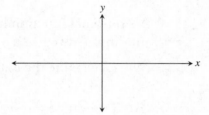

2. The graph of a sinusoidal function is shown.

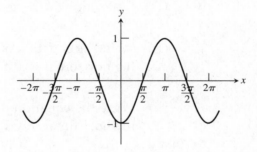

(A) Write the coordinates of two local maximum and two local minimum points.

(B) Determine the amplitude.

(C) Determine the vertical shift.

(D) Determine the period.

(E) Write the equation of a cosine function that has this graph. Identify the horizontal shift for this function.

(F) Write the equation of a sine function that has this graph. Identify the horizontal shift for this function.

Additional Practice

1. A sinusoidal function has a local maximum at $(2, 8)$ and the next local minimum at $(6, -2)$.

   (A) Determine the amplitude.

   (B) Determine the vertical shift.

   (C) Determine the range of the function.

   (D) Determine the period.

   (E) Write the equation of a cosine function that has this graph. Identify the horizontal shift for this function.

   (F) Write the equation of a sine function that has this graph. Identify the horizontal shift for this function.

   (G) Sketch the graph.

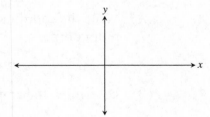

2.   A sinusoidal function has a local maximum at $(0, 2)$ and the next minimum at $(\pi/4, -2)$. A correct equation for the function is

(A) $y = 2 \cos(4x)$

(B) $y = 2 \sin(4x)$

(C) $y = 4 \cos(x)$

(D) $y = 4 \cos(4x)$

(E) $y = 4 \sin(4x)$

3.   The function $y = -3 \cos(x - \pi/4)$ has a local maximum at which point?

(A) $(\pi/4, 1)$

(B) $(\pi/4, 3)$

(C) $(3\pi/4, 3)$

(D) $(5\pi/4, 1)$

(E) $(5\pi/4, 3)$

Need More Help With . . .

Sinusoidal functions?

See . . .

Precalculus, Section 4.4

More Trigonometric Functions

 Objectives:

- Sketch the graph of a trigonometric function.
- Determine the transformations for a trigonometric function, given an equation.

Big Picture

You should be familiar with the family of functions known as *trigonometric functions*. In the previous section, we explored sine and cosine functions. This section explores the other trigonometric functions, the *tangent, cotangent, secant,* and *cosecant functions.* You should be able to sketch a graph for these types of functions, given the equation. You should also be able to describe the transformations that show how the graph compares to the basic trigonometric graph. The ability to look at these functions from an algebraic, numerical, and graphical perspective will be a great aid in understanding the concepts of precalculus and calculus.

Content and Practice

The first new trigonometric function is $y = \tan(x)$. One definition states that $\tan(x) = \dfrac{\sin(x)}{\cos(x)}$. Therefore, the domain of this function does not include all real numbers; it excludes values where $\cos(x) = 0$, namely the odd multiples of $\pi/2$. The graph of $y = \tan(x)$ will have vertical asymptotes at these values. Transformed functions may have additional coefficients. However, the main transformations are those that transform these vertical asymptotes—the horizontal stretch, shrink, or shift. The basic function $y = \tan(x)$ has period π, and the function $y = \tan(bx)$ has period $\dfrac{\pi}{|b|}$.

Each of the sine, cosine, and tangent functions has a reciprocal function. They are the cosecant, secant, and cotangent functions, respectively. All of these reciprocal functions also have vertical asymptotes. The cotangent function has period $\frac{\pi}{|b|}$, but the secant and cosecant functions have period $\frac{2\pi}{|b|}$.

1. A function f is defined as $f(x) = \tan\left(\frac{1}{2}x\right)$.

    (A) Determine the period.

    (B) Determine the domain.

    (C) Determine the equations of the vertical asymptotes.

    (D) Determine the x-intercepts.

    (E) Sketch the graph, showing several periods.

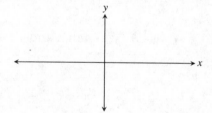

2. Write the transformations that describe how the graph of $g(x) = 2 \csc(3(x - \pi)) + 1$ compares to the graph of $f(x) = \csc(x)$.

3. Sketch each pair of functions on the same set of axes; show at least two periods and label the axes.

(A) $y = \cos x$ and $y = \sec x$

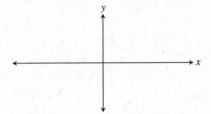

(B) $y = \sin x$ and $y = \sin\left(x - \dfrac{\pi}{2}\right) + 2$

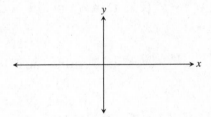

(C) $y = \tan x$ and $y = \cot x$

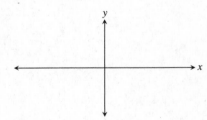

Additional Practice

1. A function f is defined as $f(x) = 3\sec(2x) + 1$.

(A) Determine the period.

(B) Determine the domain.

(C) Determine the equations of the vertical asymptotes.

(D) Determine the vertical shift.

(E) Write the coordinates of two local maximum and two local minimum points.

(F) Sketch the graph, showing several periods.

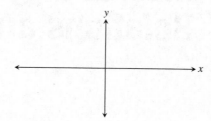

2. If the graph of $f(x) = \cot(x)$ is transformed by a horizontal shrink of $\frac{1}{4}$ and a horizontal shift left π, the result is the graph of

(A) $g(x) = \cot\left[\frac{1}{4}(x - \pi)\right]$

(B) $g(x) = \cot\left[\frac{1}{4}(x + \pi)\right]$

(C) $g(x) = \cot\left[4(x - \pi)\right]$

(D) $g(x) = \cot\left[4(x + \pi)\right]$

(E) $g(x) = \cot(4x + \pi)$

3. The function $y = \tan(x/3)$ has an x-intercept at

(A) $\pi/3$

(B) π

(C) 2π

(D) 3π

(E) 6π

Need More Help With . . .   ***See . . .***

Trigonometric functions?   *Precalculus*, Section 2.3

Inverse Trigonometric Relations and Functions

 Objectives:

• Identify the domain and range of an inverse trigonometric function.
• Compute values for inverse trigonometric relations and functions.

Big Picture

Each of the trigonometric functions has an *inverse relation*. These relations are not functions because the original functions are not one-to-one. You should be familiar with the infinite set of values for an inverse relation. However, if the range of the inverse relation is properly restricted, then the inverse is a function with only one specific value. You should be familiar with the domain and range of each of these functions.

Content and Practice

The equation $\sin(x) = \frac{1}{2}$ has an infinite number of solutions. In the *Precalculus* textbook, the symbol $\sin^{-1}\left(\frac{1}{2}\right)$ denotes a single value, $\frac{\pi}{6}$. The function $f(x) = \sin^{-1}(x)$ is the inverse sine function and it has domain $[-1, 1]$ and range $[-\pi, \pi]$. The restricted range is selected so that the inverse will have the same domain, will actually be a function, and will pass the vertical line test. We wish to accomplish the same thing with the inverse cosine but cannot use the same range because it would not pass the vertical line test. Thus, each of the trigonometric functions has an inverse. The sine, cosine, and tangent are summarized in the table below. There are certain values, such as our $\sin^{-1}\left(\frac{1}{2}\right)$ example, that you should be able to compute without a calculator; others require the use of a calculator.

Function	Domain	Range
$\sin^{-1}(x)$	$[-1, 1]$	$\left[-\frac{\pi}{2}, \frac{\pi}{2}\right]$
$\cos^{-1}(x)$	$[-1, 1]$	$[0, \pi]$
$\tan^{-1}(x)$	$(-\infty, \infty)$	$\left(-\frac{\pi}{2}, \frac{\pi}{2}\right)$

Sometimes you are asked to solve an equation such as $\sin(x) = -\frac{\sqrt{3}}{2}$. You can use the inverse trig function to find a solution. However, the equation actually has an infinite number of solutions. For example, $\sin^{-1}\left(-\frac{\sqrt{3}}{2}\right) = -\frac{\pi}{6}$, but sin is also negative in Quadrant III so $\frac{7\pi}{6}$ is also a solution. The complete set of solutions could be written as $\left\{-\frac{\pi}{6} + 2n\pi, \frac{7\pi}{6} + 2n\pi\right\}$.

1. Compute exact values.

    (A) $\cos^{-1}\left(-\frac{\sqrt{3}}{2}\right)$

    (B) $\tan^{-1}(-1)$

    (C) $\sin^{-1}\left(\frac{\sqrt{2}}{2}\right)$

2. Evaluate. (3 decimal places)

    (A) $\cos^{-1}(0.873)$

    (B) $\tan^{-1}(2.4)$

    (C) $\sin^{-1}(-0.671)$

    (D) $\sec^{-1}(-0.511)$

3. Write all solutions.

    (A) $\cos(x) = \frac{1}{2}$

    (B) $\sin(x) = -1$

    (C) $\tan(x) = -\frac{\sqrt{3}}{3}$

Additional Practice

⊞ 1. Evaluate. (3 decimal places)

(A) $\cos^{-1}(-0.246)$

(B) $\tan^{-1}(-1.5)$

(C) $\sin^{-1}(0.379)$

2. Solve each equation.

(A) $3 + \tan(x) = 2$

(B) $4\cos^2(x) = 3$

(C) $2\sin^2(x) = \sin(x)$

(D) $\cos^2(x) = 4$

3. $\sin^{-1}\left(-\dfrac{1}{2}\right) =$

(A) $-\dfrac{\pi}{6}$

(B) $-\dfrac{\pi}{3}$

(C) $\dfrac{\pi}{6}$

(D) $\dfrac{5\pi}{6}$

(E) $\dfrac{7\pi}{6}$

Need More Help With . . .

Inverse trigonometric functions?
Solving trigonometric equations?

See . . .

Precalculus, Section 4.7
Precalculus, Sections 5.1–5.4

Parametric Relations

Objectives:

- Given parametric equations, plot relations by hand or calculator.
- Control the speed and direction of the plot by varying *t* and its increments or by varying the equations.
- Produce parametric equations for Cartesian equations.
- Convert parametric equations to Cartesian equations (eliminate the parameter).
- Model motion problems.

Big Picture

Parametrics offer a powerful method to plot many relations whether or not they are functions. They also allow us to model motion, since we have more control over how points are plotted. Beyond this course, you will work with the calculus of parametrics, so gaining a high level of comfort with them now will assure future success.

Content and Practice

When you first learned to plot lines, you probably used a chart where you chose *x*-values and plugged them into an equation to produce *y*-values. With parametrics, the *x*- and *y*-values are produced independently by substituting for a third variable, *t*, called the parameter. In modeling motion, *t* usually represents time.

1. Given the following parametric equations, produce a table of values and plot the relation. The table has been started for you.

t	x	y
−2	6	−4
−1		
0		
1		
2		
3		

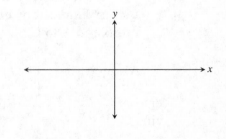

$$x_t = t^2 - t$$
$$y_t = 2t$$

2. Using substitution, convert the parametric equations in Problem 1 to Cartesian form. This is called eliminating the parameter. (*Hint:* Solve for x as a function of y.)

When parametric equations contain trigonometric functions, we often rely on a trigonometric identity rather than substitution to eliminate the parameter. Consider the parametric equations below.

$$x_t = \sec(t) \qquad y_t = \frac{1}{2}\tan^2(t)$$

If we use the trigonometric identity $1 + \tan^2\theta = \sec^2\theta$, the equation becomes $1 + 2y = x^2$.

3. Convert to Cartesian coordinates: $x = 3\sin(t)$, $y = 4\cos(t)$.
   (*Hint:* Divide each by the constant first.)

Any *function* can be converted to parametric form simply by letting the independent variable be t. So, for instance, $y = \sqrt{x^2 + 2x}$ can be converted to $x = t$ and $y = \sqrt{t^2 + 2t}$. We must realize, though, that due to limitations on the values of t we will not always produce a complete graph. For instance, $y = 2x - 1$ can be defined parametrically as $x = t$ and $y = 2t - 1$, but if t goes from -10 to 10, we would only see a plot of a segment from $(-10, -21)$ to $(10, 19)$.

4. Determine parametric equations to plot the right half of the parabola $y = (x - 2)^2$. Graph it on your calculator to see if you have achieved your goal.

$$x_t = \qquad\qquad t_{min} =$$
$$y_t = \qquad\qquad t_{max} =$$
$$\qquad\qquad\qquad t_{step} =$$

5. What is the effect of changing the increments of t, or t_{step} on the calculator? Find out by exploring. Try the following examples, comparing A to B and A to C.

(A) $x_t = 2t - 1 \qquad t_{min} = 0$
$ y_t = \ t + 1 \qquad t_{max} = 3$
$ t_{step} = 0.2$

(B) $x_t = 2t - 1$ $t_{min} = 0$
    $y_t = t + 1$ $t_{max} = 3$
                    $t_{step} = 0.02$

(C) $x_t = 2t - 1$ $t_{min} = 3$
    $y_t = t + 1$ $t_{max} = 0$
                    $t_{step} = -0.2$

(A) A compared to B: What was the effect of making the t_{step} smaller? Explain why it caused that effect.

(B) A compared to C: What was the effect of a negative t_{step}? Explain why it caused that effect.

6.   Compare the next two plots, where just the functions were changed slightly. (Make sure you plot in radians.) Explain the similarities and differences in the plots.

(D) $x_t = 3 \cos(t)$ $t_{min} = 0$
    $y_t = 3 \sin(t)$ $t_{max} = 6.3$
                        $t_{step} = 0.1$

(E) $x_t = \sin(t)$ $t_{min} = 0$
    $y_t = \cos(t)$ $t_{max} = 6.3$
                      $t_{step} = 0.1$

Parametric equations also allow us to model motion problems. We can model vertical motion, projectiles launched at an angle, circular motion, and many other kinds of motion.

To model vertical motion parametrically, let x equal any constant, e.g., $x = 1$. Let $y = -16t^2 + v_0 t + h_0$ where v_0 is the initial velocity in ft/sec and h_0 is the initial height in feet.

Projectile motion at an angle requires changing the equations slightly to $x = v_0 \cos(\theta) \cdot t$ and $y = -16t^2 + v_0 \sin(\theta) \cdot t + h_0$, where θ is the initial angle from the horizontal. These equations take into account the horizontal and vertical components of projectile motion as described in the *Precalculus* text. They ignore air resistance.

Circular motion is often modeled using $x = r \cdot \cos(b \cdot t) + c$ and $y = r \cdot \sin(b \cdot t) + d$, where r is the radius of motion, c is the horizontal shift, d is a vertical shift, and b is determined by the period.

7. A projectile is fired straight up from the ground with an initial velocity of 88 feet per second. Write parametric equations to model the motion.

8. A Ferris wheel has a diameter of 30 feet. Its lowest point is 8 feet off the ground. If it turns clockwise one full rotation each 20 seconds, write parametric equations to model a passenger's motion starting from the bottom and riding six full rotations.

Additional Practice

1. The parametric equations $x_t = 2t + 3$ and $y_t = \sqrt{t - 3}$ plot a portion of a/an
   (A) Line   (B) Parabola   (C) Circle
   (D) Ellipse   (E) Hyperbola

2. A ball is thrown with an initial velocity of 48 feet per second at an angle of 35° with the ground. If the ball is released at an initial height of 5 feet off the ground, approximately how far will it travel horizontally before striking the ground?

(A) 74 feet   (B) 68 feet   (C) 45 feet

(D) 42 feet   (E) 36 feet

Need More Help With . . .   **See . . .**

Parametric equations?   *Precalculus,* Section 6.3

Numerical Derivatives and Integrals

 Objectives:

- Estimate the slope of a curve at a particular point.
- Compute an average rate of change.
- Write an equation of the tangent line at a particular point on a particular curve.
- Estimate the area under a curve.

Big Picture

The two most fundamental concepts in all of calculus are those of a *derivative* and an *integral*. The derivative function tells us the slope of a curve at any point. A definite integral is used to compute the area under a curve.

Content and Practice

Your work with limits should have already developed the idea that places where a function looks "curved" may actually be locally linear (straight over infinitely small intervals). This allows us to talk about the slope of nonlinear functions. We define the slope of a secant on a function $f(x)$ to be

$$f(x) = \frac{f(x + h) - f(x)}{h}.$$

As the size of h gets smaller, the secant more and more accurately approximates the slope of the tangent line (if it exists) to the function at a given point $(x, f(x))$. The

$$\lim_{h \to 0} \frac{f(x + h) - f(x)}{h}$$

is the actual slope of the tangent to the function, when we can evaluate that limit.

When we cannot evaluate the limit, we can be satisfied with a fairly accurate numerical approximation we call the *numerical derivative*. Your calculator should have a built-in function to evaluate the numerical derivative. Most calculators use the symmetric difference quotient to estimate the derivative. The symmetric difference quotient uses points 0.001 units to the right and left of the place we are trying to find the tangent. Calculating the slope of the secant between those points usually provides a good approximation of the slope of the tangent:

$$m_{\tan} \approx \frac{f(x + h) - f(x - h)}{2h}, \quad \text{with } h = 0.001.$$

The particular syntax for using a numerical derivative function on your calculator may be discussed in class or can be found in your calculator manual.

Be aware that built-in numerical derivatives work only on functions. A numerical derivative must be calculated manually when only discrete data are available. Under those circumstances, we find the slope on the smallest interval containing the point whose derivative we are seeking.

We also use the limit concept to compute the area under a curve. First we partition the domain of the function into small intervals. For each interval we draw a rectangle that estimates the area under the curve. The actual height used can be chosen from the left edge, the right edge, or the center of the interval. The sum of the area of the rectangles estimates the area under the curve. Your graphing calculator should also have a built-in function to estimate this area.

1. By using $f(0.999)$ and $f(1.001)$, find an approximation of the slope of the tangent to the function $f(x) = e^{2x}$ at $x = 1$. Use the built-in numerical derivative on your calculator to verify your answer.

2. Estimate the area under $f(x) = x^2 + 1$ over $[0, 3]$. Sketch a graph and shade the appropriate area.

Additional Practice

1. Which of the following is the equation of the line tangent to $y = \frac{1}{2}x^2 + 2x$ at the point where $x = 2$?
   (A) $y = 4x$ (B) $y = 4x + 6$ (C) $y = 4x + 2$
   (D) $y = 4x - 2$ (E) No tangent can be drawn.

2. The height of an object dropped from a 200-foot building is given by $h = 200 - 16t^2$, where t is measured in seconds and h is measured in feet. What is the velocity of the object 2 seconds after it is dropped?

(A) −16 ft/sec   (B) −32 ft/sec   (C) −64 ft/sec

(D) 64 ft/sec   (E) 32 ft/sec

3. The number of gallons of water in a tub t minutes after the plug is pulled is shown in the table.

Time (minutes)	0	0.5	1	1.5	2	2.5	3	3.5	4
Gallons	120	102	88	76	65	57	50	44	40

(A) Find the average rate of change in the volume in the first 4 minutes. (Include units.)

(B) Use the data to find an estimate of how fast the volume is changing at the 2.75-minute mark. Show your work.

Need More Help With . . .

Numerical derivatives?

See . . .

Precalculus, Section 10.4

Part III

Review of AP* Calculus AB and BC Topics

To help you prepare for the AP* Calculus AB or AP* Calculus BC Examination this section is organized under the topical heading themes associated with each of those courses. Working through the respective set of AP* Objectives will help you confidently answer the essential question:

What am I expected to know and be able to do on the AP Calculus Examination?*

Analysis of Graphs

AP* Objective: Predict and explain behavior of a function
• Interplay between the geometric and analytic information

Big Picture

Analyzing graphs is a critical tool in the study of calculus. With the use of graphs we can make conjectures, solve problems, and support our written work. The graphing calculator has enabled us to quickly and easily produce graphs of functions. It is very important, however, that we have an understanding of the graphs of basic functions and their behaviors in order to determine which function best models a given situation and to select an appropriate viewing window on our calculators or axis labels on our graphs. In precalculus, your studies included the following key function behaviors: domain and range, whether a function is odd or even, symmetry, whether a function is periodic or continuous, zeros, intercepts, asymptotes, extrema, and translations.

Content and Practice

The Acorn AP* Course Description guide states that as a prerequisite to calculus, students should be familiar with the properties of the graphs of linear, polynomial, rational, exponential, logarithmic, trigonometric, inverse trigonometric, and piecewise functions. In your precalculus class, you studied twelve basic functions. These twelve are very useful for understanding graphs and their transformations.

The Identity Function

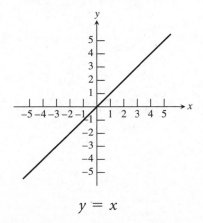

$$y = x$$

The Squaring Function

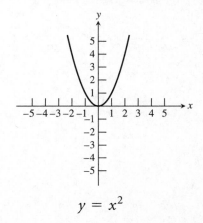

$$y = x^2$$

The Cubing Function

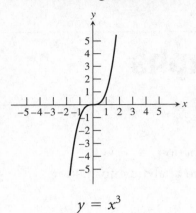

$$y = x^3$$

The Reciprocal Function

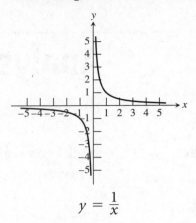

$$y = \frac{1}{x}$$

The Square Root Function

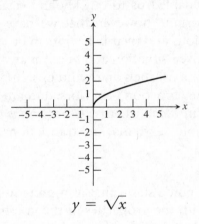

$$y = \sqrt{x}$$

The Exponential Function

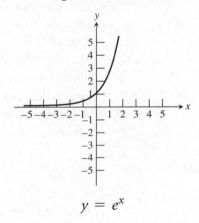

$$y = e^x$$

The Natural Logarithm Function

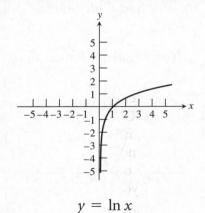

$$y = \ln x$$

The Sine Function

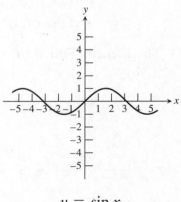

$$y = \sin x$$

The Cosine Function

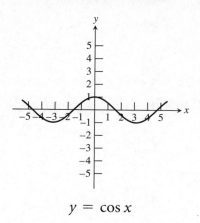

$$y = \cos x$$

The Absolute Value Function

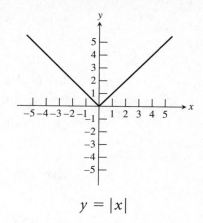

$$y = |x|$$

The Greatest Integer Function

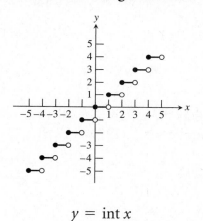

$$y = \operatorname{int} x$$

The Logistic Function

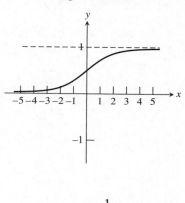

$$y = \frac{1}{1 + e^{-x}}$$

✎ *Additional Practice*

Many years ago, assessment of specific precalculus knowledge was explicitly covered within annual AP* Calculus Exams. Characteristics of functions or their graphs such as domain and range, zeros, intercepts, and symmetry were assessed in both multiple choice and open-ended questions. The following once-used AP* Free Response questions were originally created to be solved *without* the use of a graphing calculator—try your hand!

1. Let f be the real-valued function defined by $f(x) = \sqrt{1 + 6x}$. Give the domain and range of f. [1976—AB 1: Denotes the 1976 Free Response Problem 1—AP* Calculus AB Exam.]

2. Given $f(x) = x^3 - 3x^2 - 4x + 12$, find all zeros of the function f. [1976—AB 2]

3. Let $f(x) = \cos x$ for $0 \le x \le 2\pi$, let $g(x) = \ln x$ for all $x > 0$. Let S be the composition of g with f; that is, $S(x) = g(f(x))$. [1977—AB 1]

   (A) Find the domain of S.

   (B) Find the range of S.

   (C) Find the zeros of S.

4. Given the function f defined by $f(x) = x^3 - x^2 - 4x + 4$, find the zeros of f. [1978—AB 1]

5. The curve in the figure represents the graph of f, where $f(x) = x^2 - 2x$ for all real numbers x. [1979—AB 6]

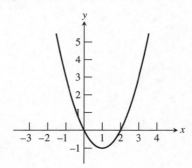

   (A) On the axes provided, sketch the graph of $y = |f(x)|$.

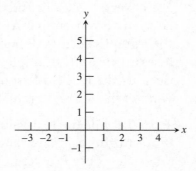

(B) On the axes provided, sketch the graph of $y = f(|x|)$.

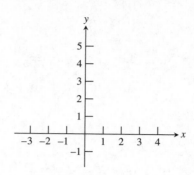

6. Let $f(x) = \ln x^2$ for $x > 0$ and $g(x) = e^{2x}$ for $x \ge 0$. Let H be the composition of f with g; that is, $H(x) = f(g(x))$, and let K be the composition of g with f; that is, $K(x) = g(f(x))$. [1980—AB 3]

(A) Find the domain of H and write an expression for $H(x)$ that does not contain the exponential function.

(B) Find the domain of K and write an expression for $K(x)$ that does not contain the exponential function.

(C) Find an expression for $f^{-1}(x)$, where f^{-1} denotes the inverse function of f, and find the domain of f^{-1}.

7. Given the function $f(x) = \cos x - \cos^2 x$ for $-\pi \le x \le \pi$. Find the x-intercepts of the graph of f. [1980—AB 5]

8. Given $f(x) = \dfrac{x^3 - x}{x^3 - 4x}$. [1982—AB 2]

(A) Find the zeros of f.

(B) Write an equation for each vertical and each horizontal asymptote to the graph of f.

(C) Describe the symmetry of the graph of f.

9. Let $f(x) = \dfrac{x + \sin x}{\cos x}$ for $-\dfrac{\pi}{2} < x < \dfrac{\pi}{2}$. State whether f is an even or an odd function. Justify your answer.   [1984—AB 2]

10. Let $f(x) = \dfrac{9x^2 - 36}{x^2 - 9}$.   [1986—AB 2]

   (A) Describe the symmetry of the graph of f.

   (B) Write an equation for each vertical and each horizontal asymptote of f.

   The following five AP* Multiple Choice questions [1969—AB] were created to be solved *without* the use of a graphing calculator.

11. Which of the following defines a function f for which $f(-x) = -f(x)$?
   (A) $f(x) = x^2$ (B) $f(x) = \sin x$ (C) $f(x) = \cos x$
   (D) $f(x) = \log x$ (E) $f(x) = e^x$
   [1969—MC 1]

12. $\ln(x - 2) < 0$ if and only if
   (A) $x < 3$ (B) $0 < x < 3$ (C) $2 < x < 3$
   (D) $x > 2$ (E) $x > 3$
   [1969—MC 2]

13. The set of all points (e^t, t), where t is a real number, is the graph of $y =$

   (A) $\dfrac{1}{e^x}$.   (B) $e^{1/x}$.   (C) $xe^{1/x}$.   (D) $\dfrac{1}{\ln x}$.   (E) $\ln x$.

   [1969—MC 10]

14. If $f(x) = \dfrac{4}{x-1}$ and $g(x) = 2x$, then the solution set of $f(g(x)) = g(f(x))$ is

(A) $\left\{\dfrac{1}{3}\right\}$.   (B) $\{2\}$.   (C) $\{3\}$.

(D) $\{-1, 2\}$.   (E) $\left\{\dfrac{1}{3}, 2\right\}$.

[1969—MC 12]

15. If the function f is defined by $f(x) = x^5 - 1$, then f^{-1}, the inverse function of f, is defined by $f^{-1}(x) =$

(A) $\dfrac{1}{\sqrt[5]{x+1}}$.   (B) $\dfrac{1}{\sqrt[5]{x+1}}$.   (C) $\sqrt[5]{x-1}$.

(D) $\sqrt[5]{x-1}$.   (E) $\sqrt[5]{x+1}$.

[1969—MC #14]

Need More Help With . . .   ***See . . .***

Functions and their properties?   *Precalculus*, Section 1.2

Calculus, Section 1.2

The twelve basic functions?   *Precalculus*, Section 1.3

Graphical transformations?   *Precalculus*, Section 1.5

Rational functions?   *Precalculus*, Section 2.7

Exponential functions?   *Calculus*, Section 1.3

Logarithmic functions?   *Precalculus*, Section 3.3

Calculus, Section 1.5

Sinusoids?   *Precalculus*, Section 4.4

Calculus, Section 1.6

Limits of Functions

Limits of functions, including one-sided limits
• An intuitive understanding of the limiting process
• Calculating limits using algebra
• Estimating limits from graphs or tables of data

Big Picture

To determine the behavior of a function, we evaluate its *limits* at significant points of its domain. We can use limits to determine where a function is continuous and where it has asymptotes, as well as to predict its values and end behavior. The concept of limits allows us to find instantaneous rates of change, a study that leads to differential calculus. We also use limits to estimate areas under curves, which leads to integral calculus. You should be able to determine limits using methods of substitution, algebra, graphing, or numerical approximation.

Content and Practice

When we are finding the limit of a function, we are determining the value that the function, $f(x)$, approaches as x gets very close to some particular value. This does not mean that $f(x)$ takes on that value at x, but rather that it *approaches* that value. Additionally, we sometimes find a limit of a function as x approaches either infinity or negative infinity (see example 5).

For a limit to exist as x approaches some value c, the limit of the function must be the same as x is approached from both the left- and right-hand sides of c. If c is either a right- or left-hand endpoint, we may be able to find a one-sided limit at that point.

A function $f(x)$ has a limit, L, as x approaches c if and only if

$$\lim_{x \to c^-} f(x) = \lim_{x \to c^+} f(x) = L.$$

We simply write $\lim_{x \to c} f(x) = L$ (if the conditions above are met).

You will notice that the definition specifies that three different conditions be met.

$$\lim_{x \to c^-} f(x) = \lim_{x \to c^+} f(x) = L$$

This is easily done by evaluating the three parts separately, as the following example illustrates.

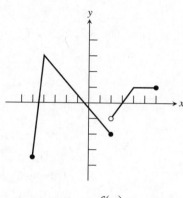

$$\lim_{x \to 2^-} f(x) = -2$$

$$\lim_{x \to 2^+} f(x) = -1$$

Since $\lim_{x \to 2^-} f(x) \ne \lim_{x \to 2^+} f(x)$,

$\lim_{x \to 2} f(x)$ does not exist.

$$y = f(x)$$
$$[-6, 6] \text{ by } [-4, 4]$$

We shall review finding two-sided and one-sided limits as well as limits that involve infinity.

Two-Sided Limits

One method used to find a limit of a function is to substitute that value into the function. Find the following limits using substitution.

1. $\lim\limits_{x \to 2} (x^3 - 5) =$

2. $\lim\limits_{x \to -4} \left(\dfrac{x^2}{x - 2} \right) =$

3.(A) $\lim\limits_{x \to 3} \left(\dfrac{x^2 - 9}{x^2 - 5x + 6} \right) =$

(B) The substitution method was not helpful in part (A). Why?

(C) Use algebra to solve this problem.

$$\lim_{x \to 3} \frac{x^2 - 9}{x^2 - 5x + 6} =$$

$$\lim_{x \to 3} \frac{(\quad)(\quad)}{(\quad)(\quad)} =$$

$$\lim_{x \to 3} \frac{\quad}{\quad} =$$

(D) Use a table of function values to approximate $\lim_{x \to 3} f(x)$.

x	$f(x)$
2.7	8.1429
2.8	7.25
2.9	6.5556
3	
3.1	5.5455
3.2	5.1667

One-sided limits

4.

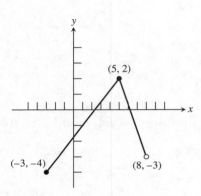

At the endpoints, we can find only the right-hand limit as x approaches -3 and the left-hand limit as x approaches 8.

(A) $\lim_{x \to -3^+} f(x) =$

(B) $\lim_{x \to 8^-} f(x) =$

Limits Involving Infinity

5.

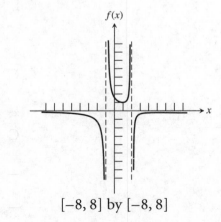

$[-8, 8]$ by $[-8, 8]$

$$f(x) = \frac{x - 3}{x^2 - x - 2}$$

(A) $\lim_{x \to -1^-} f(x) =$

(B) $\lim_{x \to -1^+} f(x) =$

(C) $\lim_{x \to 2^-} f(x) =$

(D) $\lim_{x \to 2^+} f(x) =$

(E) $\lim_{x \to -\infty} f(x) =$

(F) $\lim_{x \to +\infty} f(x) =$

Problem 5 illustrates the following properties.

> If either $\lim\limits_{x \to a^-} f(x) = \pm\infty$ or $\lim\limits_{x \to a^+} f(x) = \pm\infty$, then the line $x = a$ is a
> *vertical asymptote* of the graph of the function $y = f(x)$.

> If either $\lim\limits_{x \to +\infty} f(x) = b$ or $\lim\limits_{x \to -\infty} f(x) = b$, then the line $y = b$ is a
> *horizontal asymptote* of the graph of the function $y = f(x)$.

Additional Practice

1. $f(x) = \begin{cases} 2x - 3, & x \le 2 \\ x^2 - 1, & x > 2 \end{cases}$

    (A) $\lim\limits_{x \to 2^-} f(x) =$ (B) $\lim\limits_{x \to 2^+} f(x) =$

    (C) What does this imply about the $\lim\limits_{x \to 2} f(x)$? Explain. _____

    _____

2. Let $f(x) = \begin{cases} 2x - 3, & x \le 2 \\ x^2 + a, & x > 2. \end{cases}$

    Use one-sided limits to find the value of a so that $\lim\limits_{x \to 2} f(x) = 1$.

3.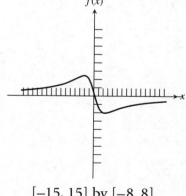

    (A) $\lim\limits_{x \to -\infty} f(x) =$

    (B) $\lim\limits_{x \to +\infty} f(x) =$

    (C) Conclusion

    _____

    _____

    $[-15, 15]$ by $[-8, 8]$

    $f(x) = \dfrac{-10x + 2}{x^2 + 5}$

4.

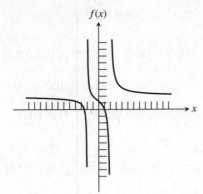

$[-12, 12]$ by $[-10, 10]$

$$f(x) = \frac{2x^2 + 3x - 5}{x^2 - 4}$$

(A) $\displaystyle\lim_{x \to -\infty} f(x) =$

(B) $\displaystyle\lim_{x \to +\infty} f(x) =$

(C) $\displaystyle\lim_{x \to -2^-} f(x) =$

(D) $\displaystyle\lim_{x \to 2} f(x) =$

(E) Conclusions

_____

_____

5.

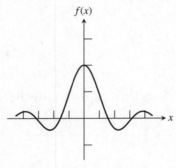

$[-4, 4]$ by $[-1, 3]$

$$f(x) = \frac{\sin 2x}{x}$$

(A) $\displaystyle\lim_{x \to 0} f(x) =$

(B) $\displaystyle\lim_{x \to +\infty} f(x) =$

6.   Use the values in the table to approximate $\displaystyle\lim_{x \to -1.8} f(x)$.

x	$f(x)$
-1.83	-22.51
-1.82	-22.54
-1.81	-22.57
-1.8	
-1.79	-22.63
-1.78	-22.66
-1.77	-22.69

7. The graph of which of the following equations has $y = 1$ as an asymptote?

(A) $y = \cos x$ 　　　　(B) $y = e^x$ 　　　　(C) $y = \dfrac{x^3}{x^2 + 1}$

(D) $y = \dfrac{x^2}{x^2 - 5}$ 　　　　(E) $y = -\ln x$

8. If $\lim\limits_{x \to a} f(x) = L$, where L is a real number, which of the following must be true?

   I. $f(a) = L$

   II. $\lim\limits_{x \to a^-} f(x) = L$

   III. $\lim\limits_{x \to a^+} f(x) = L$

(A) I only 　　　　(B) I and II 　　　　(C) I and III
(D) II and III 　　　　(E) I, II, and III

9. $\lim\limits_{x \to -\infty} \dfrac{4x^2 + x - 7}{x^2 - 5x - 3} =$

(A) 0 　　　　(B) $\dfrac{7}{3}$ 　　　　(C) 4

(D) 1 　　　　(E) Nonexistent

10. If the graph of $y = \dfrac{ax + b}{x + c}$ has a horizontal asymptote $y = -2$, a vertical asymptote $x = 4$, and an x-intercept of 1.5, then $a - b + c =$

(A) -3. 　　(B) 1. 　　(C) 5. 　　(D) -9. 　　(E) -1.

Need More Help With . . .

Limits?

See . . .

Precalculus, Section 10.3

Calculus, Sections 2.1 and 2.2

Asymptotic and Unbounded Behavior

AP* Objective:

Asymptotic and unbounded behavior
• Understanding asymptotes in terms of graphical behavior
• Describing asymptotic behavior in terms of limits involving infinity

Big Picture

Graphical models help us visualize the relationships between the variable quantities of the numerical or algebraic models. By understanding the behaviors of the graphs, we can make predictions in their real-world applications. Graphs of past and present business data can help predict future growth. Calculus studies include many different types of optimization and related rate problems, which analyze practical situations. A good knowledge of basic graphs and their behavior is invaluable in solving these problems.

Content and Practice

When we are looking for restrictions on a function's domain, we need to determine if there are any *vertical asymptotes*. Similarly, finding *horizontal asymptotes* will help us determine a function's range. The concept of boundedness of a graph also gives us insight as to the range. If we know a function's domain and range, we know on what intervals we can expect to evaluate and analyze a function.

Vertical Asymptotes

Consider the following function.

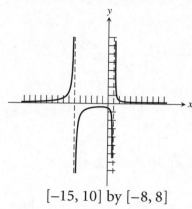

[−15, 10] by [−8, 8]

$$\lim_{x \to -6^-} f(x) = \infty \quad \text{and}$$
$$\lim_{x \to -6^+} f(x) = -\infty$$

We may conclude that the line $x = -6$ is a vertical asymptote of the graph of $f(x)$.

$$\lim_{x \to 1^-} f(x) = -\infty \quad \text{and} \quad \lim_{x \to 1^+} f(x) = \infty$$

We may conclude that the line $x = 1$ is a vertical asymptote of the graph of $f(x)$.

$$f(x) = \frac{-x + 3}{x^2 + 5x - 6}$$

> If either $\lim\limits_{x \to a^-} f(x) = \pm\infty$ or $\lim\limits_{x \to a^+} f(x) = \pm\infty$, then the line $x = a$ is a *vertical asymptote* of the graph of the function $y = f(x)$.

Horizontal Asymptotes

Looking at the graph of $f(x)$ above, we see that $\lim\limits_{x \to -\infty} f(x) = 0$ and $\lim\limits_{x \to \infty} f(x) = 0$. We may conclude that the line $y = 0$ is a horizontal asymptote of the graph of $f(x)$.

> If either $\lim\limits_{x \to +\infty} f(x) = b$ or $\lim\limits_{x \to -\infty} f(x) = b$, then the line $y = b$ is a *horizontal asymptote* of the graph of the function $y = f(x)$.

Consider the following functions.

I.

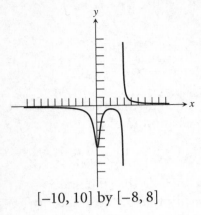

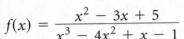

$[-10, 10]$ by $[-8, 8]$

$$f(x) = \frac{x^2 - 3x + 5}{x^3 - 4x^2 + x - 1}$$

II.

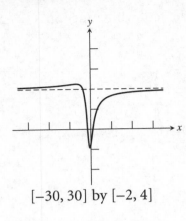

$[-30, 30]$ by $[-2, 4]$

$$g(x) = \frac{2x^2 + x - 3}{x^2 + 2x + 4}$$

III.

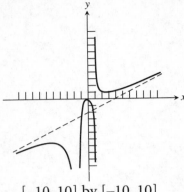

$[-10, 10]$ by $[-10, 10]$

$$h(x) = \frac{x^4 - 2x^3 + 5x^2 + 3x + 1}{2x^3 + 4x^2 - x - 5}$$

You may remember the following short cuts for determining the horizontal asymptotes of rational functions.

1. If the degree of the numerator is less than the degree of the denominator, the horizontal asymptote is $y = 0$.

2. If the degree of the numerator is equal to the degree of the denominator, the horizontal asymptote is $y = \dfrac{\text{leading coefficient of numerator}}{\text{leading coefficient of denominator}}$.

3. If the degree of the numerator is greater than the degree of the denominator, there is no horizontal asymptote.

Note that the graph of $h(x)$ above has a *slant asymptote*. A slant asymptote will be found when the degree of the numerator is 1 greater than the degree of the denominator. Using long division, $h(x)$ may be rewritten as

$$h(x) = \frac{1}{2}x - 2 + \frac{\frac{5}{2}x^2 + \frac{15}{2}x - 9}{2x^3 - 4x^2 - x - 5}.$$

For large values of x, $h(x)$ approaches the line $y = \frac{1}{2}x - 2$, which is the equation of its slant asymptote.

End Behavior Model

Suppose we want to determine the behavior of a function f where $|x|$ is very large. We use limits to define and determine the existence of an end behavior model.

A function g is a

left end behavior model for f if and only if $\displaystyle\lim_{x\to-\infty}\frac{f(x)}{g(x)}=1$.

right end behavior model for f if and only if $\displaystyle\lim_{x\to+\infty}\frac{f(x)}{g(x)}=1$.

Comparing function graphs or table values of ordered pairs can be useful in suggesting or confirming the end behavior model of a given function.

Consider the following functions.

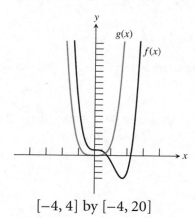

$$f(x) = 2x^4 - 5x^3 + x^2 + 1$$
$$g(x) = 2x^4$$

$[-4, 4]$ by $[-4, 20]$

For large values of x the graph of f will begin to look much like the graph of

g. Considering the limits in our end behavior definition above, we use division to express $\dfrac{f(x)}{g(x)}$ as a sum, $\dfrac{f(x)}{g(x)} = \dfrac{2x^4 - 5x^3 + x^2 + 1}{2x^4} =$

$\left(1 + \dfrac{-5x^3 + x^2 + 1}{2x^4}\right)$. It follows easily that $\displaystyle\lim_{x\to-\infty}\frac{f(x)}{g(x)} = 1$ and

$\displaystyle\lim_{x\to\infty}\frac{f(x)}{g(x)} = 1$. We therefore conclude that $g(x) = 2x^4$ is both a left end and

right end behavior model for $f(x) = 2x^4 - 5x^3 + x^2 + 1$.

Boundedness

A function f is bounded below if there is some number b that is less than or equal to every number in the range of f. Any such number b is called a lower bound of f.

A function f is bounded above if there is some number B that is greater than or equal to every number in the range of f. Any such number B is called an upper bound of f.

A function f is bounded if it is bounded both above and below.

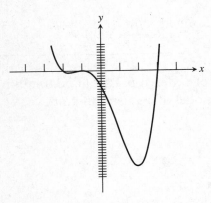

Not bounded above; bounded below

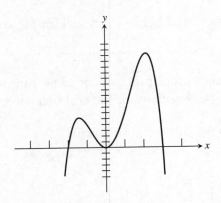

Bounded above; not bounded below

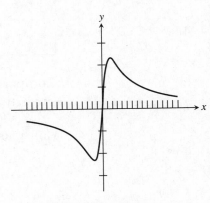

Bounded above; bounded below

Determine the equations of asymptotes for the following functions algebraically. Confirm your answers graphically.

1. $f(x) = \dfrac{x^2 + 7x + 12}{x^2 - 16}$

2. $g(x) = \dfrac{2x^2 + 3x - 5}{x - 3}$

3. $h(x) = \dfrac{4x^3}{x^4 + 1}$

4. Which of the functions in Problems 1–3 is bounded above and below?

5. Which of the following is a left end behavior model for $f(x) = x^2 - 3e^{-x}$?

(A) $y = x^2$ (B) $y = -3e^{-x}$ (C) $y = 3e^{-x}$

(D) $y = x^2 - 3$ (E) $y = e^{-x}$

✎ 1.   The function $f(x) = \dfrac{4x^2 - 3}{2x^2 + 1}$ is

   I.   unbounded.

   II. bounded below by $y = -3$.

   III. bounded above by $y = 2$.

   IV. bounded below by $y = 2$.

   (A) I   (B) IV only   (C) III only

   (D) II and III   (E) II only

✎ 2.   If $f(x) = e^x + 2$, which of the following lines is an asymptote to the graph of f?

   (A) $y = -2$ (B) $x = 0$ (C) $y = 2$

   (D) $x = 2$ (E) $y = 0$

Need More Help With . . .	See . . .
Asymptotes and end behavior models?	*Precalculus*, Sections 1.2 and 2.7 *Calculus*, Section 2.2
Boundedness?	*Precalculus*, Sections 1.2

Function Magnitudes and Their Rates of Change

Asymptotic and unbounded behavior

• Comparing relative magnitudes of functions and their rates of change (for example, contrasting exponential growth, polynomial growth, and logarithmic growth)

Big Picture

There are many functions that have values that increase as their *x*-values increase. To get a better understanding of the behavior of these functions, we often compare them to the exponential function, which grows very rapidly, or to the logarithmic function, which grows very slowly.

Content and Practice

To compare exponential, polynomial, and logarithmic functions, we shall use the following definitions.

Let $f(x)$ and $g(x)$ be positive for sufficiently large values of x.

1. f grows faster than g (and g grows slower than f) as $x \rightarrow \infty$ if

$$\lim_{x \to \infty} \frac{f(x)}{g(x)} = \infty \quad \text{or} \quad \lim_{x \to \infty} \frac{g(x)}{f(x)} = 0.$$

2. f and g grow at the same rate as $x \rightarrow \infty$ if

$$\lim_{x \to \infty} \frac{f(x)}{g(x)} = L \neq 0. \quad (L \text{ is finite})$$

Example 1: Let $f(x) = g(x) + h(x)$, where $g(x) = 5x^3$ and $h(x) = -x^2 + 3x - 6$. Using the definition above,

$$\lim_{x \to \infty} \frac{5x^3 - x^2 + 3x - 6}{5x^3} = \lim_{x \to \infty} \left(\frac{5x^3}{5x^3} + \frac{-x^2 + 3x - 6}{5x^3} \right)$$

$$= \lim_{x \to \infty} \left(1 + \frac{-x^2 + 3x - 6}{5x^3} \right)$$

$$= 1 + 0$$

$$= 1.$$

From this we see that $f(x)$ and $g(x)$ grow at the same rate. This is why, for large values of x, we can ignore the terms of $h(x)$ in $f(x)$. This is also why we can say that $g(x)$ is a right end behavior model of $f(x)$.

Example 2: Compare the growth rates of x^5 and e^x as $x \to \infty$.

$$\lim_{x \to \infty} \frac{x^5}{e^x} = \lim_{x \to \infty} \frac{5x^4}{e^x} \qquad \text{Use L'Hôpital's Rule.}$$

$$= \lim_{x \to \infty} \frac{20x^3}{e^x}$$

$$= \lim_{x \to \infty} \frac{60x^2}{e^x}$$

$$= \lim_{x \to \infty} \frac{120x}{e^x}$$

$$= \lim_{x \to \infty} \frac{120}{e^x}$$

$$= 0$$

Therefore, x^5 grows slower than e^x as $x \to \infty$.

Example 3: Compare the growth rates of $\log \sqrt[3]{x}$ and $\ln x$ as $x \to \infty$.

$$\lim_{x \to \infty} \frac{\log \sqrt[3]{x}}{\ln x} = \lim_{x \to \infty} \frac{\frac{\ln x}{3 \ln 10}}{\ln x}$$

$$= \frac{1}{3 \ln 10}$$

Therefore $\log \sqrt[3]{x}$ and $\ln x$ grow at the same rate as $x \to \infty$.

1. List the functions e^x, 3^x, and x^3 in order from slowest-growing to fastest-growing as $x \to \infty$.

   (A) $e^x, 3^x, x^3$ (B) $3^x, x^3, e^x$ (C) $x^3, e^x, 3^x$

   (D) $x^3, 3^x, e^x$ (E) $e^x, x^3, 3^x$

2. Which of the following functions grow at the same rate as $x \to \infty$?

   I. $f(x) = x^3$

   II. $g(x) = \sqrt{x^6 + x^2}$

   III. $h(x) = \sqrt[3]{x^6 + 9x^3}$

   IV. $j(x) = \dfrac{x^5 - 4x^2 + 3}{x^2 + 2x - 9}$

   (A) I and II only   (B) I and IV only

   (C) I, II, and IV only   (D) I, II, and III only

   (E) II, III, and IV only

Need More Help With . . .   *See . . .*

Relative rates of growth?   *Calculus,* Section 8.3

Continuity

AP* Objective: Continuity as a property of functions
- An intuitive understanding of continuity
- Understand continuity in term of limits

Big Picture

In precalculus, your first experience with *continuity* was studying functions and their properties. This study included finding various limits of a function over its domain. Continuity is presented early in the *Calculus* text since most theorems in the course rely on this property. Remember, a function can be examined for continuity at a point, on an interval, or over its entire domain.

Content and Practice

Intuitively we think of a function as continuous over an interval if its graph can be sketched in one continuous motion without lifting the pencil. In studying functions that are discontinuous, we can learn more about the behaviors of continuous functions.

By considering the graph, identify all points of discontinuity in each of the following.

1. $f(x) = \tan x$ 2. $g(x) = \dfrac{x^2 - 2x - 3}{x + 1}$

3. $h(x) = \begin{cases} 2x - 3, & x \le -1 \\ x^2 - 5, & x > -1 \end{cases}$

In your own words, write a sentence to explain how you identified the discontinuities of each function in Problems 1–3.

4. $f(x)$ _____

5. $g(x)$ _____

6. $h(x)$ _____

The *Calculus* book presents the following definition of continuity at an interior point of a domain.

A function $y = f(x)$ is continuous at an interior point c of its domain if
$$\lim_{x \to c} f(x) = f(c)$$
If c is an endpoint of its domain, only the appropriate one-sided limit is checked.

Many students find it helpful to recognize that this definition asks three distinct questions:

   I.   Does $f(c)$ exist? (What do I get at $x = c$?)

   II.   Does $\lim_{x \to c} f(x)$ exist? (What do I expect to get as x approaches c?)

   III. Does $\lim_{x \to c} f(x) = f(c)$? (Is what I get at $x = c$ equal to what I expected to get as x approaches c?)

Look again at the functions in Problems 1–3. Determine which part of the definition of continuity is not satisfied in each.

7. $f(x)$ _____

8. $g(x)$ _____

9. $h(x)$ _____

Additional Practice

In Problems 1–3, use the definition of continuity to decide whether each of the following functions is continuous at the specified value of x. If it is not continuous, explain why the function does not meet the definition.

1. $f(x) = \lfloor x \rfloor$ at $x = 3$

2. $g(x) = \begin{cases} x + 5, & x \neq 0 \\ 4, & x = 0 \end{cases}$ at $x = 0$

3. $h(x) = \begin{cases} -x^2 + 8, & x < 2 \\ \frac{1}{2}x + 3, & x \geq 2 \end{cases}$ at $x = 2$

4. Let f be the function defined as follows:

$$f(x) = \begin{cases} |x - 3| + 1, & x < 3 \\ ax^2 + bx, & x \geq 3 \end{cases}$$

(A) If $a = 3$ and $b = 2$, is f continuous for all x? Justify your answer.

(B) Describe all values of a and b for which f is a continuous function.

5. Which of the following functions are continuous for all real numbers x?
    I. $f(x) = |x|$
    II. $f(x) = \tan x$
    III. $f(x) = 3x^2 + x - 7$

    (A) I only   (B) II only   (C) III only
    (D) I and II   (E) I and III

6. Let $f(x) = \dfrac{x^3 - 2x^2 - 29x - 42}{x^2 - 9}$. Which of the following statements is true?

(A) $f(x)$ has a removable discontinuity at $x = -3$.

(B) $f(x)$ has a jump discontinuity at $x = 3$.

(C) If $f(3) = \dfrac{-5}{3}$, then $f(x)$ is continuous at $x = 3$.

(D) $f(x)$ has nonremovable discontinuities at $x = -3$ and $x = 3$.

(E) $\displaystyle\lim_{x \to -3} f(x) = \infty$

Need More Help With . . .

Continuity?

See . . .

Precalculus, Section 1.2

Calculus, Section 2.3

Intermediate and Extreme Value Theorems

AP* Objective:

Continuity as a property of functions
- Geometric understanding of graphs of continuous functions (Intermediate Value Theorem and Extreme Value Theorem)

Big Picture

Considerable time is spent in algebra II and precalculus courses learning how to find the zeros of a function and then using this information to sketch its graph. As you extended your mathematical knowledge in calculus, you learned that the zeros of the graph of a function's derivative will enable us to determine a function's maximum or minimum values. These are values we need when solving optimization problems.

In precalculus, you used synthetic division to determine if a number was a zero of a polynomial function. The Intermediate Value Theorem helps us determine where such zeros exist. The Extreme Value Theorem gives us insight as to whether a function has maxima or minima. It is important to remember that the extreme values are the maximum or minimum *y*-values of the function.

Content and Practice

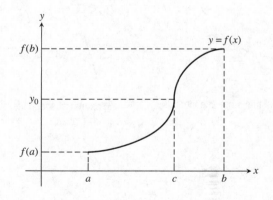

The Intermediate Value Theorem

A function $y = f(x)$ that is continuous on a closed interval $[a, b]$ takes on every value between $f(a)$ and $f(b)$ on (a, b).

If y_0 is between $f(a)$ and $f(b)$, then $y_0 = f(c)$ for some c in (a, b).

It is essential that f be a continuous function in order to apply the Intermediate Value Theorem, as illustrated below.

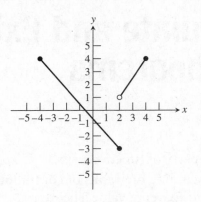

$$f(x) = \begin{cases} -x - 1, & -4 \le x \le 2 \\ \dfrac{3}{2}x - 2, & 2 < x \le 4 \end{cases}$$

We see that $f(1) = -2$ and $f(4) = 4$. However, $f(x)$ does not take on all values between -2 and 4 on the interval $[1, 4]$. This is because $f(x)$ is not a continuous function on the interval $[1, 4]$.

However, looking at the interval $[-4, 2]$ where f is continuous, we see that $f(x)$ does take on every value between $f(-4)$ and $f(2)$.

The Extreme Value Theorem

If f is continuous on a closed interval $[a, b]$, then f has both a maximum value and a minimum value on the interval.

Maxima and minima can occur at interior points or at the endpoints, as illustrated in the figures.

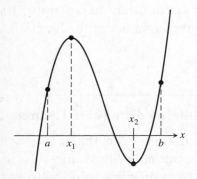

Maximum and minimum at interior points

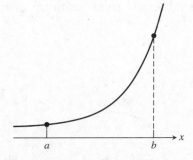

Maximum and minimum at endpoints

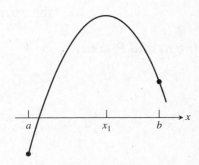

Minimum at endpoint; maximum at interior point

We can see that f must be continuous on a closed interval in order to apply the Extreme Value Theorem by analyzing the following graph.

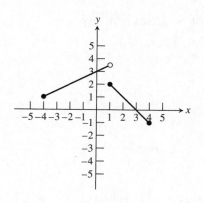

$$f(x) = \begin{cases} \frac{1}{2}x + 3, & -4 \le x < 1 \\ -x + 3, & 1 \le x \le 4 \end{cases}$$

On the interval $[-4, 4]$, there is no maximum value for $f(x)$. There is also no maximum on $[-4, 1)$ since this is not a closed interval. There is a maximum on $[-4, 0.9]$.

Additional Practice

Use the graph of f to the right for Problems 1 and 2.

1. Explain how the Intermediate Value Theorem is used to verify that f has a zero between $x = 2$ and $x = 3$.

2. Approximate the maximum and minimum values of f on the interval $[0, 2]$.

3. The function f is continuous on the closed interval $[-2, 1]$. Some values of f are shown in the table.

x	-2	-1	0	1
$f(x)$	-3	7	k	3

The equation $f(x) = \frac{3}{2}$ must have at least two solutions in the interval $[-1, 1]$ if $k =$

(A) 1.　　(B) $\frac{3}{2}$.　　(C) 2.　　(D) $\frac{5}{2}$.　　(E) 3.

4.   A function f is continuous on $[-4, 1]$ and has its maximum at $(-3, 5)$ and its minimum at $\left(\frac{1}{2}, -6\right)$. Which of the following statements must be false?

(A) The graph of f crosses both axes.
(B) f is always decreasing on $[-4, 1]$.
(C) $f(-2) = 0$
(D) $f(-1) = 6$
(E) $f(0) = 2$

5.   Let $f(x) = \left| \cos(x) - \frac{1}{2} \right|$. Which is the maximum value attained by f?

(A) $\frac{1}{2}$ (B) 1 (C) $\frac{3}{2}$ (D) π (E) 2π

Need More Help With . . .	*See* . . .
Intermediate Value Theorem?	*Precalculus,* Section 2.3
	Calculus, Section 2.3
Extreme Value Theorem?	*Calculus,* Section 4.1

Parametric, Polar, and Vector Functions

AP* Objective:

Parametric, polar, and vector functions

Big Picture

In function mode, y is a function of the independent variable x.

$$y = f(x)$$

In parametric mode, x and y are both functions of the independent parameter t.

$$x = f(t)$$
$$y = g(t)$$

In polar mode, r is a function of the independent variable θ.

$$r = f(\theta)$$

Vector-valued functions utilize parametric equations.

$$\mathbf{r}(t) = <f(t), g(t)> \text{ with alternative notation}$$
$$\mathbf{r}(t) = f(t)\mathbf{i} + g(t)\mathbf{j}$$

Content and Practice

Parametric equations are often used to describe the motion of a particle in the plane. A graphing calculator can be used to see the path and direction of the particle.

1. During the time period from $t = 0$ to $t = 6$ seconds, a particle moves along the path given by

    $$x(t) = 4 \cos \pi t$$
    $$y(t) = 5 \sin \pi t$$

    (A) Find the position of the particle when $t = 2.5$.

(B) Sketch the graph of the path of the particle from $t = 0$ to $t = 6$. Indicate the direction of the particle along its path.

(C) How many times does the particle pass through the point found in part (a)?

A calculator may be used on some problems to graph polar functions, but take care when using a graph to solve a system of polar equations.

2.   Solve the following system.

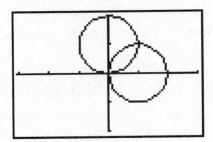

$$r = 2 \sin \theta$$
$$r = 2 \cos \theta$$

From the graph there appear to be two solutions: $(0, 0)$ and $\left(\frac{\pi}{4}, \sqrt{2} \right)$. However, substituting the ordered pairs in the system of equations shows only the second pair is a solution. This can be seen graphically if the calculator is placed in simultaneous mode before the graphs are drawn.

Vector-valued functions can be handled on the calculator using parametric mode.

3.   The position of a moving particle is given by the vector function

$$\mathbf{r}(t) = <\cos(\pi t), t - 1> \text{ with alternate notation}$$
$$\mathbf{r}(t) = \cos(\pi t)\mathbf{i} + (t - 1)\mathbf{j}$$

(A) Find the position vector for the particle at $t = 1.1$

(B) Graph the path of the particle for $0 \le t \le 2$.

Additional Practice

1. The position of a particle in a plane is described by the vector-valued function $r(t) = \langle e^{-t}, \cos t \rangle$. What is the position of the particle at $t = \pi/2$?

(A) $(e^{-\pi/2}, 0)$ (B) $(e^{-\pi/2}, 1)$ (C) $(-e^{-\pi/2}, 0)$

(D) $(-e^{-\pi/2}, 1)$ (E) $\left(0, \dfrac{\pi}{2}\right)$

2. Consider the following polar functions.

$$r_1 = 4\sin\theta$$
$$r_2 = 2$$

(A) Graph the functions.

(B) Find the points of intersection of the graphs of r_1 and r_2.

3. A particle moves along the path specified by the following parametric equations.

$$x(t) = \sin 2t$$
$$y(t) = \cos 2t$$

Sketch the path of the particle.

Need More Help With . . .

    Parametric equations?

    Vector-valued functions?

    Polar functions?

See . . .

Precalculus, Section 6.3

Calculus, Section 10.1

Calculus, Section 10.2

Calculus, Section 10.3

Concept of the Derivative

AP* Objective:

Concept of the derivative
- Derivative presented graphically, numerically, and analytically
- Derivative defined as an instantaneous rate of change
- Derivative defined as the limit of the difference quotient

Big Picture

The concept of the *derivative* is a critical part of almost everything we do in calculus; it is important to familiarize ourselves with the derivative as seen from different perspectives: graphically, numerically, and analytically. This familiarity will in turn allow us to understand and apply the derivative in a variety of situations.

Content and Practice

From a graphical perspective, a function's derivative at any specific point can be thought of as the slope of the graph of that function at that point. Shown at right is the graph of a quadratic function (in bold) and of its derivative. Notice that high positive or negative derivative values indicate rapid growth or decline in the function and a steeply sloping graph, whereas derivative values close to zero indicate little or no change in the values of the original function and an approximately horizontal graph.

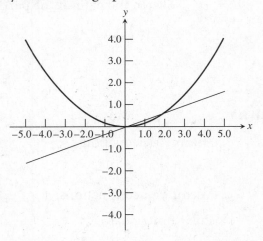

1. Consider the graph shown at right.

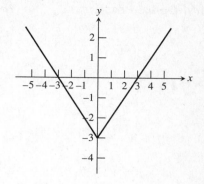

   (A) Give the value of the derivative of this function at $x = -3$.

   (B) Give the value of the derivative of this function at $x = 1$.

Similarly, the derivative of a function can be thought of as the rate at which the function's value is changing at a specific instant. For example, if $f(t)$ measures the position of a moving particle at time t, then $f'(5)$ represents the velocity of that particle at the moment when $t = 5$.

2. If $y = f(x)$ is a profit function measuring the amount of profit (in dollars) as a result of manufacturing and selling x basketballs, what is the significance of $f'(550)$? Make sure you use specific units.

Since we generally require two points to calculate slope, the task of finding slope at a single point will require a new strategy. We will use the standard slope formula $\dfrac{f(b) - f(a)}{b - a}$ for two points on the curve, and then take the limit of this difference quotient as one point approaches the other.

Eventually, we will be able to develop analytical techniques for finding the derivative as a *function* related to the original function.

1. Let $y = g(x)$ be a function that measures the water depth in a pool x minutes after the pool begins to fill. Then $g'(25)$ represents:

   I. The rate at which the depth is increasing 25 minutes after the pool starts to fill.

   II. The average rate at which the depth changes over the first 25 minutes.

   III. The slope of the graph of g at the point where $x = 25$.

   (A) I only   (B) II only   (C) III only

   (D) I and II   (E) I and III   (F) I, II, and III

2. The function $y = f(x)$ measures the fish population in Blue Lake at time x, where x is measured in years since January 1, 1950. If $f'(25) = 500$, it means that

   (A) there are 500 fish in the lake in 1975.

   (B) there are 500 more fish in 1975 than there were in 1950.

   (C) on the average, the fish population increased by 500 per year over the first 25 years following 1950.

   (D) on January 1, 1975, the fish population was growing at a rate of 500 fish per year.

   (E) none of the above.

Need More Help With . . .   *See . . .*

       Concept of derivative?   *Precalculus,* Section 10.1

                                               Calculus, Section 2.4–3.6

Differentiability and Continuity

AP* Objective:

Concept of the derivative
• Relationship between differentiability and continuity

Big Picture

Most, but not all, of the functions we encounter in calculus will be differentiable over their entire domains. Before we can confidently apply the rules regarding derivatives, we need to be able to recognize the exceptions to the rule.

Content and Practice

A function that is differentiable at a point or over an interval will always be continuous there, but the converse is not true: There are situations where a continuous function may not have a derivative. To rephrase this, a function that is *dis*continuous at a point will definitely *not* have a derivative at that point. A continuous function, on the other hand, will still fail to have a derivative at any point where it has a corner, a cusp, or a vertical tangent.

1. Consider the function shown at right. At what domain values does the function appear to be

   (A) differentiable?

   (B) continuous but not differentiable?

   (C) neither continuous nor differentiable?

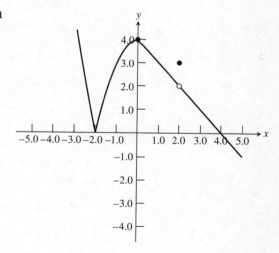

1. Let f be a function with $f'(5) = 8$. Which of the following statements is true?

    (A) f must be continuous at $x = 5$.

    (B) f is definitely not continuous at $x = 5$.

    (C) There is not enough information to determine whether or not $f(x)$ is continuous at $x = 5$.

2. Consider the function $y = f(x)$ shown at right.

    (A) At what x-values is f discontinuous?

    (B) At what x-values would this function *not* be differentiable?

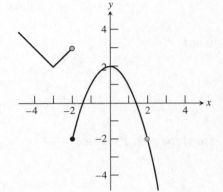

3. Suppose f is a function such that $f'(9)$ is undefined. Which of the following statements is true?

    (A) f must be continuous at $x = 9$.

    (B) f is definitely not continuous at $x = 9$.

    (C) There is not enough information to determine whether or not f is continuous at $x = 9$.

4. Suppose that f is a function that is continuous at $x = -11$. Which of the following statements is true?

    (A) f must be differentiable at $x = -11$.

    (B) f is definitely not differentiable at $x = -11$.

    (C) There is not enough information to determine whether or not $f(x)$ is differentiable at $x = -11$.

5. Which of the following statements are always true?

    I. A function that is continuous at $x = c$ must be differentiable at $x = c$.

    II. A function that is differentiable at $x = c$ must be continuous at $x = c$.

    III. A function that is *not* continuous at $x = c$ must *not* be differentiable at $x = c$.

    IV. A function that is *not* differentiable at $x = c$ must *not* be continuous at $x = c$.

(A) None of them   (B) I and III   (C) II and IV

(D) I and IV   (E) II and III   (F) I, II, III, and IV

Need More Help With . . .

    Differentiability?

    Continuity and differentiability?

See . . .

    Precalculus, Section 10.1

    Calculus, Section 3.2

Slope of a Curve at a Point

AP* Objective:

Derivative at a point

• Slope of a curve at a point. Examples emphasized include points at which there are vertical tangents and points at which there are no tangents.

Big Picture

We often use graphs to capture the relationship between variables (distance versus time, for instance). We know from previous courses that when we do this for a linear function, the constant slope of that function represents the **rate of change** of one variable with respect to the other. By defining slope for nonlinear functions, we can extend this same concept to a much broader range of situations.

Content and Practice

The value of a function's derivative at a specific point can be thought of as the slope of the tangent line to the function's graph at that point. This interpretation can be quite helpful, both as a means of approximating the value of a derivative and as a means for identifying points at which the derivative will be undefined.

1.   Estimate the slope of each curve at point *P*.

(A)

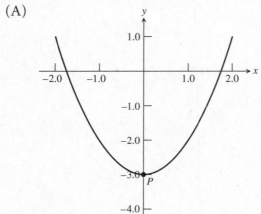

(B)

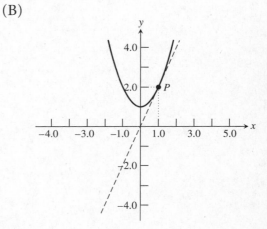

(C)

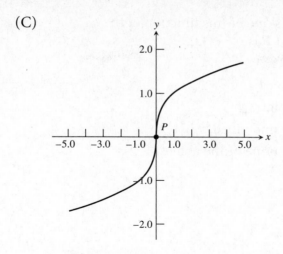

Although the conventional formula for slope requires two distinct points, the slope of the tangent line to a function at a specific point can be determined by finding the slope between that point and a nearby point on the curve, and then finding the limit as the nearby point approaches the original point.

> The **slope** of the curve $y = f(x)$ at the point $(a, f(a))$ is
> $$\lim_{h \to 0} \frac{f(a + h) - f(a)}{h},$$
> if and only if it exists.

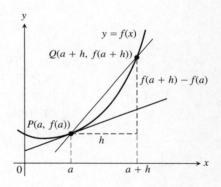

The tangent slope is

$$\lim_{h \to 0} \frac{f(a + h) - f(a)}{h}.$$

2. Consider the function $f(x) = x^2 + 3$.

    (A) Using your calculator to view an appropriate graph of the function, estimate the slope of the curve at $x = 3$.

    (B) Justify your answer analytically, using the definition of the slope of a curve.

3. The slope of this function at the point P is

(A) 1.

(B) −1.

(C) 0.

(D) undefined.

(E) none of the above.

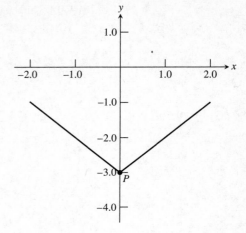

4. Explain how your answer to Problem 3 is consistent with the definition of the slope of a curve.

From the above examples, we can see that there are at least two situations in which a function's slope may not be defined at a certain point: the tangent line to the function at that point may be vertical (and therefore have no slope), or there may simply be no tangent line at that point.

Additional Practice

 1. Find all values of x at which the slope of the function $f(x) = \frac{1}{x}$ is equal to $-\frac{1}{4}$.

Questions 2 and 3 refer to the function $y = f(x)$ shown in the figure.

2. At what x-value(s) within the domain of the graph is the slope of f approximately zero?
   (A) Never
   (B) $\{-1, 1\}$
   (C) $\{-1, 0, 1\}$
   (D) $\{-0.45, 0.45\}$
   (E) $\{0\}$

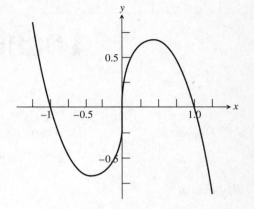

3. At what x-values within the domain of the graph would the slope of f be undefined?
   (A) Never   (B) $\{-1, 1\}$ (C) $\{-1, 0, 1\}$
   (D) $\{-0.45, 0.45\}$ (E) $\{0\}$

Need More Help With . . .

    Derivative at a point?

See . . .

Precalculus, Section 10.1

Calculus, Section 2.4

Local Linearity

AP* Objective:

Derivative at a point

• Tangent line to a curve at a point and local linear approximations

Big Picture

You first encountered the derivative $y = f'(x)$ as the limiting slope of a secant to the curve as the second point of intersection approached the first. As those points converged, the secant line became a tangent to the curve. For differentiable functions, that tangent line can be used as a simple approximation, or model, for the curve near the tangent point. This is a significant practical application; it is also helpful in developing an intuitive understanding of what makes a curve differentiable.

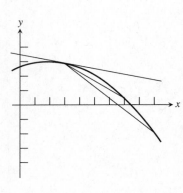

Content and Practice

The derivative of a function at a specific point, $f'(a)$, gives us the slope of a tangent line to the function f at the point where $x = a$. We can construct the equation of that tangent line from the values of $f(a)$ and $f'(a)$.

1.   Write the equation of the tangent line to the f at $x = 5$, given that $f(5) = 3$ and $f'(5) = 0.5$.

   This tangent line can be used as a model to find approximate values for the original function close to the point of tangency.

2.   Use your tangent line from Problem 1 to approximate $f(5.023)$.

Why is this approximation close to the value of the original function? Because the derivative is defined as the (two-sided) limit of the slope at that point, the derivative can only exist at a point if the limiting slopes on both sides of that point are identical—if there is any kind of corner or cusp at that point, it will *not* be differentiable!

Consider the function $f(x) = |x| + 1$, which is not differentiable at $x = 0$, and $g(x) = \sqrt{x^2 + 0.0001} + 0.99$, which *is* differentiable at $x = 0$. In many viewing windows, this pair of functions will appear almost identical around $x = 0$. If we zoom in, as shown below, and consider only a very small interval surrounding $x = 0$, we can see that the differentiable function g flattens out and begins to appear linear over that interval. The nondifferentiable function f maintains its sharp corner even at this resolution.

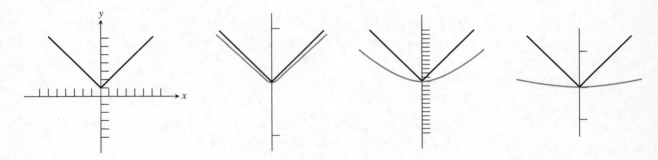

As this example illustrates, a differentiable function—even one that appears to be changing direction suddenly—will begin to look and act like its tangent line if we zoom in close enough and look at a very small interval around the point of tangency. This characteristic of differentiable functions is known as *local linearity*.

Additional Practice

1. Let $f(x) = e^x$. Write an equation for the tangent line to this function at $x = 2$.

2. For each of the following functions,

    I. State whether or not it is differentiable at $x = 1$. (How do you know?)

    II. If the function is differentiable, give an equation for the line tangent to the function at $x = 1$.

    (A) $f(x) = x^2 + 1$

    (B) $f(x) = 2x$

    (C) $f(x) = \begin{cases} x^2 + 1 & x \le 1 \\ 2x & x > 1 \end{cases}$

3. Consider the function $f(x) = \sin(kx) + 3$. Given that $f'(0) = k$, what is the approximate value of $f(0.03)$?

    (A) $3k + 0.03$ (B) 3.03 (C) $3.03k$

    (D) $k + 3.03$ (E) $0.03k + 3$

4. Consider the function $f(x) = a \ln(x + 2)$. Given that $f'(1) = \frac{a}{3}$, what is the *approximate* value of $f(0.98)$?

    (A) $\left(\dfrac{a}{3}\right) \cdot (-0.02) + a \cdot \ln(2.98)$

    (B) $\dfrac{-0.02a}{3}$

    (C) $(0.98) \cdot \ln\left(\dfrac{a}{3}\right)$

    (D) $\left(\dfrac{a}{3}\right) \cdot a \cdot \ln(2.98) + 0.98$

    (E) $\left(\dfrac{a}{3}\right) \cdot 0.98 + a \cdot \ln(2.98)$

***Need More Help With* . . .**   *See* . . .

    Derivatives?   *Precalculus,* Section 10.1

    Derivatives and local linearity?   *Calculus,* Sections 3.1 and 3.2

Instantaneous Rate of Change

Derivative at a point
• Instantaneous rate of change as the limit of average rate of change
• Approximate rate of change from graphs and tables of values

Big Picture

Many of our applications are concerned with *rates of change*. While the difference quotient gives us a way to measure the average rate of change over an interval, often we are more interested in the rate of change at one given instant. The use of limits will allow us to find such a rate.

Content and Practice

Suppose we have a particle that is moving in such a way that its position at time t is given by the function $f(t) = 3t^2$, with t measured in seconds. Also let us assume that we are interested in calculating the rate at which the particle is moving at the instant $t = 1$. We can calculate the **average** rate of change of $f(t)$ over any interval using the difference quotient $\dfrac{f(t + h) - f(t)}{h}$, where h is the length of the interval. For instance, the average rate of change for the 10-second period beginning at $t = 1$ would be

$$\text{average rate} = \frac{f(1 + 10) - f(1)}{10}$$

$$= \frac{f(11) - f(1)}{10}$$

$$= \frac{363 - 3}{10}$$

$$= 36 \text{ units/sec}$$

1. Repeat the above calculation for the (A) 5-second, (B) 3-second, and (C) 1-second intervals beginning at $t = 1$. What do you think the rate of change is at precisely $t = 1$?

The **instantaneous rate of change** of $f(x)$ at the moment when $x = a$ can be calculated by finding the average rate of change of $f(x)$ over the small interval between $x = a$ and a nearby value $x = a + h$ and then taking the limit as this interval is made increasingly smaller.

$$\text{Instantaneous Rate of Change} = \lim_{h \to 0} \frac{f(a + h) - f(a)}{h}.$$

2. Calculate the instantaneous rate of change for $a = 1$ analytically, using the above definition. (*Hint:* simplify the numerator until you can factor out h.) How does this result compare to your result in Problem 1?

When we are working from a graph or table, it is not always possible to evaluate the function at any point we choose. It is still possible, however, to approximate the value of an instantaneous rate of change at a point by calculating the average rate of change over a small interval including that point.

3. Find an approximate value for the rate of change of $f(x)$ at $x = 3$ based on the information in the table. Choose the most appropriate value.

x	$f(x)$
0	5.75
2	6.15
4	8.32
6	12.66

(A) 7.235
(B) 2.170
(C) 1.085
(D) 0.645
(E) 0.200

Additional Practice

1. Let $f(x) = 6 - x^2$. Use analytical methods to find the instantaneous rate of change when $x = -2$.

2. Consider the following table of values for the advertising budget of ACME Cola.

Year	Budget
1994	28.9
1996	33.3
1998	35.5
2000	39.0
2002	44.8
2004	54.5

(A) Find the average rate of change for the period 1994–2004.

(B) Find the average rate of change for the period 1996–2002.

(C) Find the average rate of change for the period 1998–2000.

(D) Give an estimate for the instantaneous rate of change in 1999.

3. Consider the function $y = f(x)$ shown at right. Approximate the instantaneous rate of change at $x = 2$.

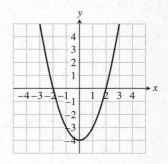

(A) -4

(B) -2

(C) 0

(D) 2

(E) 4

Need More Help With . . .   ***See . . .***

Instantaneous velocity?   *Precalculus*, Section 10.1

Instantaneous rate of change?   *Calculus*, Sections 2.4 and 3.4

Relationships between the Graphs of *f* and *f'*

Derivative as a function
- Corresponding characteristics of graphs of *f* and *f'*
- Relationship between the increasing and decreasing behavior of *f* and the sign of *f'*

Big Picture

A large part of what we study in calculus hinges on the relationship between a function and its derivative. One important aspect of this relationship is the connection between the graphs of the two functions—the graph of *f* can give us important information about the graph of *f'*, and vice versa. A basic knowledge of what each graph can reveal about the other is critical in order to understand many of the theorems you will encounter in calculus. It is also an important tool for applications such as optimization and modeling.

Content and Practice

At any given point, the value of a function's derivative can be thought of as the slope of that function. Therefore, if we know what the graph of a specific function *f* looks like, we can get a good idea of what its derivative function *f'* looks like simply by estimating the slope of *f* at various points along the graph and plotting each slope at its corresponding *x*-value. After plotting a number of these points, a smooth curve can be drawn through the points to approximate the graph of the derivative function *f'*.

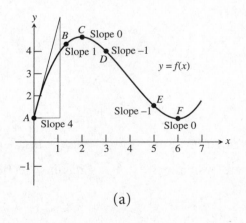

(a)

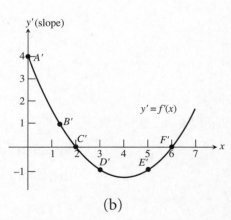

(b)

1. What do you notice about the values of the derivative function on those intervals where the original function *f* shown above is

    (A) increasing?

    (B) decreasing?

2. Explain why these observations make sense in terms of the slope of *f*.

3. When the graph of *f'* crosses the *x*-axis, what does this tell you about the graph of *f*? Explain, in terms of slope, why this happens.

Additional Practice

Questions 1–3 relate to the function *f* shown in the figure

1. Give the approximate value(s) of the *x*-intercepts of *f'*.

    (A) *f'* has no *x*-intercepts.

    (B) {3}

    (C) {−1, 1, 3}

    (D) {0, 2}

    (E) {−3, 3}

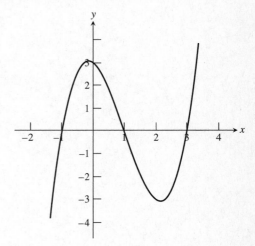

2. Over what interval(s) will the graph of *f'* have negative values?

    (A) (−∞, ∞)   (B) (−∞, 0) ∪ (2, ∞)

    (C) (0, 2)   (D) (−∞, −1) ∪ (1, 3)

    (E) Never

3. Based on the appearance of their graphs, which of these functions looks like it could be equal to its own derivative?

(A) $f(x) = \sin x$ (B) $f(x) = \cos x$ (C) $f(x) = e^x - 5$

(D) $f(x) = e^{x-5}$ (E) None of the above

Need More Help With . . .

    Concept of derivative?

    Relationship between graphs of f and f'?

See . . .

    Precalculus, Section 10.1

    Calculus, Sections 3.1 and 3.2

The Mean Value Theorem

AP* Objective:

Derivative as a function
• The Mean Value Theorem and its geometric consequences

Big Picture

One consequence of the Mean Value Theorem is that it provides a method of determining the intervals where the graph of a function rises or falls. This skill is useful in many situations where we have a model that we wish to analyze (for instance, we can examine a position function and determine at what times the object was moving to the left or the right).

Content and Practice

The statement of the Mean Value Theorem is as follows:

Mean Value Theorem for Derivatives

If $y = f(x)$ is continuous at every point of the closed interval $[a, b]$

and differentiable at every point of its interior (a, b), then there

is at least one point c at which

$$f'(c) = \frac{f(b) - f(a)}{b - a}.$$

This theorem says that, under these conditions, there must be at least one point in the interval where the instantaneous rate of change, $f'(c)$, equals the average rate of change, $\frac{f(b) - f(a)}{b - a}$, for the interval.

Graphically speaking, if A and B are two points on a differentiable curve, then somewhere between points A and B there is at least one tangent line to the curve that is parallel to chord AB.

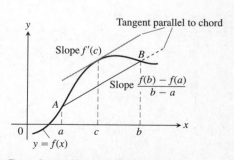

Figure for the Mean Value Theorem

115

1.   Consider the function $f(x) = x^2$ on the interval $[1, 5]$.

(A) Are the conditions of the Mean Value Theorem met?

(B) Find $f'(x)$ as a function in terms of x.

(C) Find the average rate of change $\dfrac{f(b) - f(a)}{b - a}$ over the interval $(a = 1$ and $b = 5)$.

(D) Use your answers to parts (b) and (c) to find each value of c in the interval that satisfies the equation of the Mean Value Theorem, $f'(c) = \dfrac{f(b) - f(a)}{b - a}$.

One important consequence of the Mean Value Theorem is that it simplifies the process of determining where a graph rises or falls.

Corollary

Let f be continuous on $[a, b]$ and differentiable on (a, b).
1.   If $f' > 0$ at each point of (a, b), then f increases on $[a, b]$.
2.   If $f' < 0$ at each point of (a, b), then f decreases on $[a, b]$.

Two other important conclusions can be drawn from the Mean Value Theorem:

I.   If $f' = 0$ over an interval, then f is constant over that interval.

II. If $f' = g'$ over an interval, then $f = g + C$ over the interval for some constant C.

1. Consider the function $f(x) = \sqrt{x-2}$. On what intervals are the hypotheses of the Mean Value Theorem satisfied?

   (A) $[0, 2]$ (B) $[1, 5]$ (C) $[2, 7]$ (D) None of the these

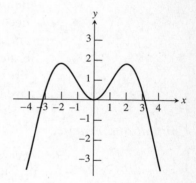 2. Verify that the function $f(x) = \sin x$ satisfies the hypotheses of the Mean Value Theorem on the interval $[2, 11]$. Then approximate to 3 decimal places all values of c in $(2, 11)$ that satisfy the Mean Value Theorem equation.

3. Consider the following graph of $f(x) = x \sin x$ on the domain $[-4, 4]$. How many values of c in $(-4, 4)$ appear to satisfy the Mean Value Theorem equation?

   (A) None

   (B) One

   (C) Two

   (D) Three

   (E) Four or more

Need More Help With . . .

Average rate of change?

Mean Value Theorem?

See . . .

Precalculus, Sections 2.1 and 10.1

Calculus, Section 4.2

Equations Involving Derivatives

Derivative as a function
• Equations involving derivatives. Verbal descriptions are translated into
  equations involving derivatives, and vice versa.

Big Picture

Being able to write and comprehend basic equations involving derivatives is a
necessary prerequisite for later applications, including related rate problems
and solving differential equations. Fluency with such equations also makes it
easier to recognize and deal with exponential growth and decay situations,
since one of the characteristics of such a situation is the fact that the derivative (growth rate) is proportional to the function value itself.

Content and Practice

We know that the derivative $f'(x)$ measures the rate at which the quantity represented by $f(x)$ changes with respect to change in x. Likewise, dy/dx measures
the rate at which y changes with respect to the change in x. This information
can help us to write and interpret statements involving derivatives. For
instance, suppose that we have a function $T(x)$ that measures surface temperature of an object (x is time, measured in hours). If we are told that the object's
surface is cooling off at a steady 5°C per hour, then we can write $T'(x) = -5$.

1. The water level $W(t)$ (where t is measured in hours) is falling 3 inches
    every hour. Write an equation involving a derivative to describe the sit-
    uation.

2. The rabbit population $R(t)$ (where t is measured in years) increases by 10 percent every year. Which of the following equations is consistent with that description?

(A) $R'(t) = 10$

(B) $R'(t) = -10$

(C) $R'(t) = 0.10R$

(D) $R'(t) = 10R$

(E) None of the above

It is equally important to be able to translate such statements in the reverse direction; when you are presented with an equation involving derivatives, you should be able to interpret what information this gives you about the problem situation.

3. A plane is descending for a landing. Its altitude $A(t)$ is measured in feet with t measuring time in minutes. Given that $A'(t) = -200$, describe the plane's behavior.

Additional Practice

1. At time $t = 5$, the rate at which the volume of a sphere $V(t)$ is increasing is numerically equal to 8 times the rate at which its radius $r(t)$ is increasing. Write an equation to match this statement.

2. The cost of operating the widget factory $C(w)$ increases \$23 for every widget produced. Write an equation involving a derivative to describe the situation.

3. Let $A(t)$ represent the deer population in a local forest preserve at time t years, when $t \geq 0$. The population is increasing at a rate directly proportional to $1200 - A(t)$, where the constant of proportionality is k. Which of the following statements accurately reflects the situation?

(A) $A(t) = k[1200 - A(t)]$

(B) $A(t) = k[1200 - A'(t)]$

(C) $A'(t) = k[1200 - A(t)]$

(D) $A'(t) = [1200 - kA(t)]$

(E) $A'(t) = \dfrac{[1200 - A(t)]}{k}$

Need More Help With . . .

Derivative equations?

See . . .

Calculus, Sections 3.3 and 4.6

Correspondences among the Graphs of f, f', and f''

AP* Objective:

Second derivatives
• Corresponding characteristics of the graphs of f, f', and f''
• Relationship between the concavity of f and the sign of f''

Big Picture

We will often use the derivative and second derivative functions to give us information about the graph of f, and vice versa. The relationships between these functions allow us to perform optimization (where we seek to find the maximum or minimum value of a function).

Content and Practice

A function's derivative measures the slope of that function. This simple fact can lead to a number of connections between the graph of f and f' (for now, let's assume that f is a continuous function).

1. Complete each of the following statements for a continuous function f:

    (A) When f' is positive, it means that f is

    _____ .

    (B) When f' is negative, it means that f is

    _____ .

    (C) When f' is changing from negative to positive, f is at a

    _____ value.

    (D) When f' is changing from positive to negative, f is at a

    _____ value.

In the figure at right, the bold function is f and the other function is f'. Note that f' is negative on the interval $(-2, 2)$ (where the graph of f is decreasing) and positive on both $(-\infty, -2)$ and $(2, \infty)$ (where the graph of f is increasing). Points where the first derivative is changing sign correspond to maximum or minimum values of the original function.

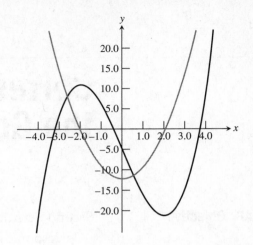

Since f'' measures the slope of f', we can apply the same basic reasoning again. When f'' is positive, for instance, it means that f' is increasing. This, in turn, means that the graph of f must be concave up.

Shown at right is the same function f (again in bold) along with its second derivative, f''. Note that $f''(x)$ is negative on $(-\infty, 0)$, where the original function is concave down; similarly, it is positive on $(0, \infty)$, where the original function is concave up. Points where the second derivative has a change of sign represent points of inflection in the original function if a tangent exists at that point.

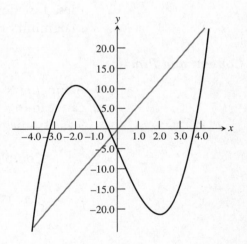

1. Consider the graph of $f(x) = x \sin x$ shown at right. Draw schematic diagrams to approximate where the first and second derivatives are positive, negative, and zero.

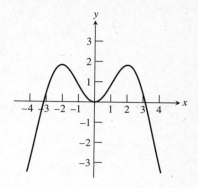

(A) f' ⟵——————————⟶

(B) f'' ⟵——————————⟶

(C) State the reasoning behind your diagrams in parts (A) and (B). How did you determine where each function would be positive, negative, or zero? Be specific.

2. Let the graph shown in Problem 1 be g'.

(A) Estimate the intervals on which $g(x)$ is increasing.

(B) Estimate the intervals on which $g(x)$ is decreasing.

(C) Estimate where $g(x)$ has local extreme values.

3. f is continuous on $[0, 8]$ and satisfies the following:

x	$0 \le x < 3$	3	$3 < x < 5$	5	$5 < x < 6$	6	$6 < x \le 8$
f''	$-$	0	$+$	Does not exist	$-$	0	$-$

(A) Based on this information, is there a point of inflection at $x = 3$?
(a) Definitely
(b) Possibly
(c) Definitely not

(B) Based on this information, is there a point of inflection at $x = 5$?

    (a) Definitely

    (b) Possibly

    (c) Definitely not

(C) Based on this information, is there a point of inflection at $x = 6$?

    (a) Definitely

    (b) Possibly

    (c) Definitely not

4.   Shown at right is a graph of f' (in bold) and f''. Sketch a possible graph of f on the same axes.

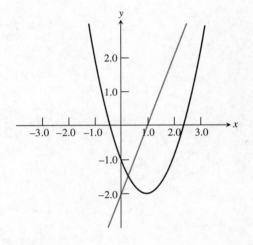

Need More Help With . . .

    Functions and their properties?

    Graphical relationships among f, f', and f''?

See . . .

Precalculus, Section 1.2

Calculus, Section 4.3

Points of Inflection

AP* Objective:

Second derivatives
• Points of inflection as places where concavity changes

Big Picture

We are interested in *points of inflection* because they help us to describe the behavior of a graph (along with the extrema of the function). They also represent points of significant change in the quantity being measured as well as its rate of change. For this reason, it can be quite useful to locate the points of inflection in some applications (it can represent, for instance, a shift in demand or price movement).

Content and Practice

A point where the graph of a function has a tangent line and where the concavity changes is a **point of inflection.** Points of inflection can be found by looking for changes in sign in the second derivative. Since the Intermediate Value Theorem applies here, the second derivative will either have a value of zero or be undefined at the points of inflection.

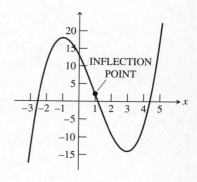

$$f(x) = x^3 - 3x^2 - 9x + 13$$

To identify points of inflection:
  I.   Locate the x-values at which the second derivative function is zero or undefined.
  II. Test values in between those points to determine the sign of the second derivative for that interval.
  III. At those points where the sign changes, if the original function has a tangent line, then that is a point of inflection.

1.   Consider the function $f(x) = -x^4 + 4x^3 + 3x + 5$.

(A) Find the second derivative, f''.

(B) At what x-values is f'' zero or undefined?

(C) Over what intervals is f'' positive? Negative?

(D) Identify any points of inflection.

Additional Practice

1.   A function $f(x)$ exists such that $f''(x) = (x - 2)^2(x + 1)$. How many points of inflection does $f(x)$ have?
(A) None   (B) One   (C) Two
(D) Three   (E) Cannot be determined

2.   Find all points of inflection of the function
$f(x) = 2x^3 - 3x^2 + 6x - 10$.

3.   Suppose f is continuous on $[0, 6]$ and satisfies the following:

x	0	3	5	6
f	-1	4	-1	-3
f'	5	0	-8	0
f''	-1	-3	Does not exist.	3

x	$0 < x < 3$	$3 < x < 5$	$5 < x < 6$
f'	$+$	$-$	$-$
f''	$-$	$-$	$+$

(A) Identify all points of inflection.

    (a) There are no points of inflection.

    (b) $(5, -1)$ only

    (c) $(3, 4)$ and $(6, -3)$

    (d) $(3, 4)$ only

    (e) $(3, 4), (6, -3)$, and $(5, -1)$

(B) Explain the reason for your choice in part (A).

Need More Help With . . .   *See* . . .

    Points of inflection?   *Calculus*, Section 4.3

Concavity of Functions

Applications of derivatives
• Analysis of curves, including the notions of monotonicity and concavity

Big Picture

The second major component of function analysis is concavity, which is related to the second derivative of a function. Concavity also represents the rate of change of the slope of a given function. Visually we see this displayed in the curvature of the function. The second derivative is also important for locating inflection points and, in combination with the first derivative, may be used to justify local extrema. Common student errors in finding inflection points are to overlook places where the second derivative is undefined—and to fail to check for changes in the sign of the second derivative.

Content and Practice

Questions on this topic can be graphical, analytic, numerical, or conceptual, as the following sample questions show.

1. Using the capital letters below the x-axis, identify the intervals where the plotted function, f is concave down. Convention dictates that intervals of concavity are reported as *open* intervals.

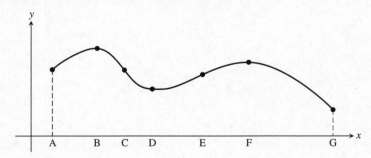

2. Given that the graph below is a plot of h' on the domain $[A, G]$, identify all places where h has points of inflection on the given domain. Explain your choice(s).

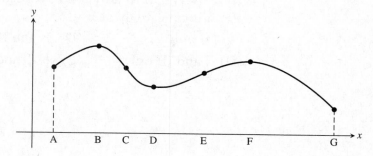

3. Is it possible for $f''(x)$ to equal 0 at $x = c$ but not have an inflection point at $x = c$? Explain.

4. Is it possible for $f''(x)$ to fail to exist at $x = a$ but still have an inflection point at $x = a$? Explain.

5. Without a calculator, find the intervals where $g(x) = \ln(4 + x^2)$ is concave down. Show work that leads to your answer.

6. Let f'' be a continuous and monotonic decreasing function on the domain $[1, 7]$. The table below shows four function values of f''.

x	1	3	5	7
$f''(x)$	5	2	1	-2

Which of the following statements must be true?

    I.   If $f'(2) = 11$, then $f'(3) > 11$.

    II. f has an inflection point between $x = 5$ and $x = 7$.

    III. f is concave up at $x = 6$.

    (A) II only   (B) II and III only   (C) I and II only

    (D) I and III only   (E) I, II, and III

Additional Practice

1.   The graph of g, a twice-differentiable function, is shown below. Choose the correct order for the values of $g(1)$, $g'(1)$, and $g''(1)$.

    (A) $g(1) < g'(1) < g''(1)$

    (B) $g'(1) < g''(1) < g(1)$

    (C) $g''(1) < g(1) < g'(1)$

    (D) $g'(1) < g(1) < g''(1)$

    (E) Cannot be determined

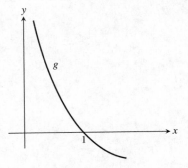

2.   Use the second derivative test for local extrema to justify that $h(x) = x^2 e^x$ has a local maximum at $x = -2$.

Need More Help With . . .	See . . .
Polynomial function behavior?	*Precalculus*, Section 2.3
Derivatives?	*Precalculus*, Sections 3.3 and 3.5–3.9
Concavity?	*Calculus*, Section 4.3

Extreme Values of Functions

AP* Objective:

Applications of derivatives
• Analysis of curves, including the notions of monotonicity and concavity

Big Picture

Determining *maximum and minimum values* on open and closed intervals is a foundational skill that applies to many other topics such as optimization and motion. The places we search for maximum and minimum values are at critical points and endpoints of closed intervals. Critical points are where the first derivative of a function is zero or where the first derivative does not exist. Later we extend the idea to relate extrema to second derivatives. Three frequently overlooked aspects of optimization are checking places where the derivative is undefined, checking endpoints, and justifying that a point is indeed a maximum or minimum by looking for a sign change in the derivative or using the second derivative test. Another important distinction is the difference between local and absolute extrema. Mastering this concept on simple functions and graphs will contribute greatly to your success on later topics.

Content and Practice

A key skill is the ability to distinguish between *local* (or relative) maximum or minimum values and *absolute* extrema. Whenever you work with a continuous function on a closed interval, endpoints must be examined as possible local and absolute extrema. Questions on this topic can be graphical, analytic, numerical, or conceptual, as the following sample questions show.

1. Identify each point on the curve as a local maximum or minimum, absolute maximum or minimum, both a local and an absolute, or neither.

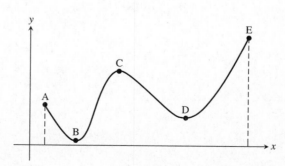

2. Is it possible for $f'(x)$ to equal 0 at $x = c$ but f not have a local maximum or minimum at $x = c$? Explain.

3. Is it possible for $f'(x)$ to fail to exist at $x = a$ but f still have a maximum or minimum at $x = a$? Explain.

4. Without a calculator, find the absolute maximum value of

   $g(x) = \frac{1}{3}x^3 - 4x$ on the interval $[-1, 4]$. Justify your answer.

5. Let f be a continuous, differentiable, and monotonic function on the domain $[3, 8]$. The table shows four function values of f.

x	3	4	6	7
$f(x)$	-4	1	5	8

Which of the following statements must be true?

    I. $f(8) > 9$

    II. $f'(5) > 0$

    III. $f'(c) = 3$ for exactly one c in $[3, 7]$

(A) II only   (B) II and III only   (C) III only

(D) I and III only   (E) I, II, and III

1. Let g be a function defined and continuous on the closed interval $[a, b]$. If g has a local minimum at c where $a < c < b$, which of the following statements must be true?

    I.   If $g'(c)$ exists, then $g'(c) = 0$.

    II. $g(c) < g(b)$

    III. g is monotonic on $[a, b]$.

    (A) I only   (B) II only   (C) III only

    (D) I and II only   (E) I and III only

2. On the interval $[-5, 5]$, f is continuous and differentiable. If $f'(x) = (x - 1)(2x + 1)(x + 3)^2$, briefly explain the following conclusions.

    (A) There is a local maximum on f at $x = -\frac{1}{2}$.

    (B) There is a horizontal tangent but no extrema at $x = -3$.

    (C) If $f(2) = 7$, then $f(3) > 7$.

Need More Help With . . .   *See . . .*

Polynomial function behavior?   *Precalculus,* Section 2.3

Taking derivatives?   *Precalculus,* Sections 3.3 and 3.5–3.9

Extreme values of functions?   *Calculus,* Section 4.1

Analysis of Parametric, Polar, and Vector Curves

AP* Objective: Applications of derivatives

• Analysis of planar curves given in parametric form, polar form, and vector form, including velocity and acceleration vectors

Big Picture

If the position of a particle in the plane is given by the vector/parametric function

$$\mathbf{r}(t) = <x(t), y(t)> \text{ with alternate notation}$$
$$\mathbf{r}(t) = x(t)\mathbf{i} + y(t)\mathbf{j},$$

then the velocity and acceleration vectors are

$$\mathbf{v}(t) = <x'(t), y'(t)> \quad \text{with alternate notation} \quad \mathbf{v}(t) = x'(t)\mathbf{i} + y'(t)\mathbf{j}$$
$$\mathbf{a}(t) = <x''(t), y''(t)> \quad\quad\quad\quad\quad\quad\quad\quad \mathbf{a}(t) = x''(t)\mathbf{i} + y''(t)\mathbf{j}$$

The slope of the path of the particle is $\dfrac{dy}{dx} = \dfrac{dy/dt}{dx/dt}$.

A polar curve of the form $r = f(\theta)$, can be defined paramaetrically by

$$x = r\cos(\theta) = f(\theta)\cos(\theta)$$
$$y = r\sin(\theta) = f(\theta)\sin(\theta)$$

Using the slope of the path of a particle for a paramatic curve it is easier to derive, rather than memorize, the equation for the slope of a tangent to a polar curve. Instead of dy/dt and dx/dt we find $dy/d\theta$ and $dx/d\theta$ remembering to use product rule. The result is

$$\frac{dy}{dx} = \frac{f'(\theta)\sin(\theta) + f(\theta)\cos(\theta)}{f'(\theta)\cos(\theta) - f(\theta)\sin(\theta)}.$$

Content and Practice

1.   A particle moves along a curve defined by the vector function
$$\mathbf{r}(t) = <(t^2 - 1), \sin(2t)>$$

   (A) Find the velocity vector.

   (B) Find the acceleration vector.

(C) Find the slope of the path of the particle at $t = \pi/6$.

2. A particle moves along a planar curve according to the parametric
    equations
    $$x(t) = t^3 - t$$
    $$y(t) = (2t - 1)^3 \quad \text{where } t \geq 0.$$

    (A) What is the velocity vector at $t = 1$?

    (B) Find the acceleration vector of the particle when the particle's
         vertical position is 0.

3. At which values of ztheta in the internal $[0, 2\pi]$ does the polar curve
    $r = 1 + \cos\theta$ have horizontal tangents?

Additional Practice

1. The position of a particle is given by $r(t) = <2t, \ln t>$.
    (A) Find the acceleration vector of the particle.

    (B) Find the speed of the particle at $t = \dfrac{1}{2}$.

2. For what values of t does the curve given by the parametric equations
    $$x(t) = \frac{t^3}{3} + \frac{t^2}{2} - 6t + 1$$
    $$y(t) = t^2 + t + 1$$

    have a vertical tangent?
    (A) 2 only   (B) -3 and 2 only   (C) -3, 0 and 2

    (D) $\dfrac{1}{2}$ (E) No value

3. A particle moves along the parametric curve
$$x(t) = t^3 - t$$
$$y(t) = t^2$$, where $0 \leq t \leq 3$.

For what values of t is the particle moving to the left?

(A) $(0, 3)$ (B) $\left(0, \sqrt{3}\right)$ (C) $\left(\sqrt{3}, 3\right)$

(D) $\left[0, 1/\sqrt{3}\right)$ (E) $\left[1/\sqrt{3}, 3\right)$

Need More Help With . . .	See . . .
Position, velocity, and acceleration vectors?	*Calculus*, Section 10.2
Slopes of polar curves?	*Calculus*, Section 10.3

Optimization

AP* Objective:

Applications of derivatives
• Optimization, both absolute (global) and relative (local) extrema

Big Picture

Optimization is one of the many applications of derivatives of functions. They are most commonly presented as word problems and encompass a wide variety of interesting applications. It is important to develop mastery of the *process* of solving this type of problem and not to try to memorize a small set of problem types.

Content and Practice

In precalculus you solved optimization problems by setting up equations and using a calculator to determine a maximum or minimum value of the function to be optimized. Since the AP* Calculus exam does not expect or allow students to use that feature of a calculator, you must be able to set the first derivative equal to zero and justify extreme values using calculus methods. The strategy for solving max-min problems is presented in Section 4.4 of the *Calculus* text.

Let's look at an example and connect it to steps from the strategy.

Example: A rectangle is inscribed under the graph of $h(x) = 9 - x^2$. Find the maximum possible area for that rectangle.

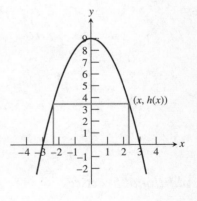

Understand the problem:

Recognize that the area of the rectangle changes based on the placement of the corners. Find the area of the largest possible rectangle.

Develop a mathematical model:

The total width of the rectangle is $2x$ since it is symmetric to the y-axis. The height of the rectangle is determined by the function value on the parabola. Therefore the area as a function of x is $A(x) = 2x(9 - x^2)$.

Identify the critical points and endpoints:

Expanding $A(x) = 18x - 2x^3$, so $A'(x) = 18 - 6x^2$. The zeros of $A'(x)$ are $x = \pm\sqrt{3}$. The only candidates for x to produce a maximum area are the endpoints $x = 0$ and $x = 3$, and the positive solution to $A'(x) = 0$, which is $x = \sqrt{3}$.

Solve the mathematical model:

The values $x = 0$ and $x = 3$ make the area of the rectangle equal to 0. Using a signed number line for $A'(x)$ shows $x = \sqrt{3}$ is a location of a maximum on A. Justification also requires a brief explanation of the signed number line.

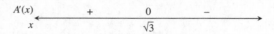

Since the first derivative changes from positive to the left of $x = \sqrt{3}$, to zero at $x = \sqrt{3}$, and to negative on the right of $x = \sqrt{3}$, a local maximum of the area function occurs at $x = \sqrt{3}$.

Interpret the solution:

Stopping at $x = \sqrt{3}$ is a very common mistake. The problem asks for the *maximum area*. The area of the rectangle is $A(x) = 2x(9 - x^2)$, so the maximum area is $A(\sqrt{3}) = 12\sqrt{3}$.

Sometimes a problem requires writing two equations. One is usually based on a fixed quantity and is used for substitution purposes. The other equation is frequently the quantity we wish to optimize. Try the next one yourself using the hint for guidance.

1. An open-topped box with square base must be constructed with a volume of 12 cubic inches. What dimensions use the least amount of material?

    (*Hint:* Define your variables for dimensions of the box. Now write two equations, one for volume and one for surface area. Which one will be optimized?)

Additional Practice

1. A particle moves from $(0, 1)$ to the right, following the path of $f(x) = e^{x/2}$. What are the coordinates of the particle when its distance to the point $(2, 0)$ is at a minimum?

    (A) $(0.841, 1.523)$ (B) $(0.841, 1.914)$ (C) $(1.415, 1.411)$

    (D) $(1.415, 2.030)$ (E) $(2, e)$

2. From an 8 inch by 10 inch rectangular sheet of paper, squares of equal size will be cut from each corner. The flaps will then be folded up to form an open-topped box. Find the maximum possible volume of the box.

3. The steps of the optimization process are reworded slightly and listed in random order below. Number them from 1 to 8 as they should be completed.

    ____ Justify that your solution provides a maximum or minimum.

    ____ Find the derivative of the varying quantity.

    ____ If necessary, substitute from the fixed quantity into the varying quantity.

    ____ Make sure you have answered the original question, with appropriate units of measure.

    ____ Define variables to be used.

    ____ Read and understand the problem, noting especially what is to be optimized.

    ____ Find the zeros of the derivative equation.

    ____ Write equations for all fixed and varying quantities.

Need More Help With . . .	*See . . .*
Writing mathematical models?	*Precalculus,* Section 1.6
Taking derivatives?	*Precalculus,* Sections 3.3 and 3.5–3.9
Optimization?	*Calculus,* Section 4.4

Related Rates

AP* Objective:

Applications of derivatives
• Modeling rates of change, including related rate problems

Big Picture

Related rates explore a relationship among several derivatives. Related rates is one of the more challenging topics during the first half of a calculus course. A methodical approach will provide the best opportunity for success. It is likely you will need practice to master this concept. Related rates may show up on either the Multiple Choice or the Free Response section of the AP* Calculus exam. Carefully defining variables and summarizing a lot of preliminary information will keep you organized.

Content and Practice

The goal of the primary technique for doing related rate problems is to identify a mathematical relationship between the quantities that are varying. It is also crucial to understand which quantities are changing and which are constant throughout the problem. Express rates as derivatives and quantities as simple variables. In certain cases, there is a secondary equation from which a substitution of variables is made prior to the differentiation process. The basic process is: write an equation relating variables, differentiate with respect to time, substitute, and solve.

1.   Air is being blown into a sphere at a rate of 6 cubic inches per minute. How fast is the radius changing when the radius of the sphere is 2 inches? Follow the steps below to review the process.

      (A) Identify the primary function to use: $V_{\text{sphere}} = \frac{4}{3}\pi r^3$.

(B) Identify or calculate all quantities at the moment in question.

"Cubic inches per minute" is a volume rate, so $\frac{dV}{dt} = 6$ in^3/min. The radius, r, is changing and therefore must be substituted *after* differentiation. If the radius is 2 inches, the volume is $\frac{4}{3}\pi 2^3 = \frac{32\pi}{3}$ cubic inches. It too is changing and must be substituted *after* differentiation.

(C) Identify the quantity being sought.

"How fast is the radius changing ..." means we want dr/dt.

(D) Differentiate both sides of the volume equation with respect to t.

$$\frac{dV}{dt} = \frac{4}{3}\pi 3r^2 \frac{dr}{dt}$$

An alternate method is to differentiate with respect to the independent variable and then use the Chain Rule.

$$\frac{dV}{dr} = \frac{4}{3}\pi 3r^2$$
$$\frac{dV}{dt} = \frac{dV}{dr}\frac{dr}{dt}$$

Substituting from the first equation into the second produces

$$\frac{dV}{dt} = \frac{4}{3}\pi 3r^2 \frac{dr}{dt}.$$

(E) Substitute known quantities at the moment in question.

$$6\frac{\text{in.}^3}{\text{min}} = \frac{4}{3}\pi 3(2\text{ in.})^2 \frac{dr}{dt}$$

(F) Solve for the desired quantity, including units whenever they exist in a problem.

$$\frac{dr}{dt} = \frac{3}{8\pi}\frac{\text{in.}}{\text{min}}$$

Follow this procedure for the following problems.

2. The edge of a cube is increasing at a rate of 2 inches per minute. At the instant the edge is 3 inches, how fast is the volume increasing?

3.   A point moves along the curve $y = (x - 3)^2$ such that its x-coordinate is increasing at 4 units per second.

(A) At the moment $x = 1$, how fast is its y-coordinate changing? Interpret your answer based on the shape of the graph and the location of the point.

(B) At the moment $x = 1$, how fast is the point's distance from the origin changing?

In part (B) of the previous problem you may have encountered three rates, the change in x and y, and the distance from the origin. The next problem requires working with three rates. Be aware of your need for the Product Rule when differentiating.

4.   Two ships leave a port at the same time, traveling on paths that differ by 50°. The first ship holds a steady course at 35 miles per hour. The second ship holds its course going 28 miles per hour. After two hours, how fast is the distance between the ships increasing? (*Hint:* Use the Law of Cosines: $a^2 = w^2 + h^2 - 2wh \cos \theta$.)

Often when a cone is involved in a problem, a second equation must be used to complete the problem. Work the following problem using the provided hint.

5.   Water is flowing into an inverted right circular cone at a rate of 4 cubic inches per minute. The cone is 16 inches tall and its base has a radius of 4 inches. At the moment the water has a depth of 5 inches, how fast is the radius at the surface of the water increasing?

(*Hint:* The volume of a cone is $V = \frac{1}{3} \pi r^2 h$, but based on proportions in the cone, substitute an expression for h so your equation is in terms of only V and r prior to differentiating.)

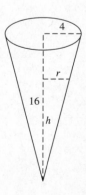

Additional Practice

1. The diagonal of a square is increasing at a rate of 3 inches per minute. When the area of the square is 18 square inches, how fast (in inches per minute) is the perimeter increasing?

   (A) 18 (B) $3\sqrt{2}$ (C) $\dfrac{3\sqrt{2}}{2}$

   (D) 6 (E) $6\sqrt{2}$

2. A spherical snowball with diameter 4 inches is removed from the freezer in June and begins melting uniformly such that it is shrinking 2 cubic inches per minute. How fast (in square inches per minute) is its surface area decreasing when the radius is 1 inch?

   (A) $\dfrac{1}{2\pi}$ (B) $\dfrac{1}{\pi}$ (C) 1

   (D) 4 (E) $4\pi^2$

Need More Help With . . .   *See . . .*

Related rates?   *Calculus,* Sections 4.6

Implicit Differentiation

AP* Objective: Applications of derivatives
                            • Use implicit differentiation to find the derivative of an inverse function

Big Picture

There are situations where it is not possible to analytically isolate y as a function of x. One example of this is $y^3 + \sin y = \cos x + x^2$. In this case we must find dy/dx using *implicit differentiation*. One of its most important uses is to help establish the formulas for the derivatives of inverse trigonometric functions. Although you are expected to know the formulas themselves the actual derivation of those formulas has rarely, if ever, appeared on the AP* exam. Review Section 3.8 of the *Calculus* text for examples of the derivations.

Content and Practice

There are three key skills for successfully differentiating implicitly with respect to x:

1. Every time you differentiate a y-term you get a dy/dx as part of the derivative.

2. Terms with xy or x/y require the Product Rule or Quotient Rule.

3. Terms with y^n require using the Power Rule with the Chain Rule, producing dy/dx.

When first learning differentiation, students often use a very fundamental skill for successful implicit differentiation without realizing it. To find the derivative of a simple explicit function such as $y = x^3$, we naturally write $dy/dx = 3x^2$. Our attention is often on the right-hand side of that equation, but let's focus on the left-hand side for a minute. You have actually differentiated the y-term with respect to x. It would be no different if the equation had started out as $y - x^3 = 0$. There are 3 "terms" to differentiate: x^3, 0, and y. When you differentiate y *with respect to x*, you get the dy/dx term, producing $dy/dx - 3x^2 = 0$. We do not see anything after $3x^2$ because we are differentiating with respect to x, so in effect we get $3x^2\, dx/dx = 3x^2 \cdot 1$.

The steps for finding dy/dx implicitly can be summarized as follows:

(A) Differentiate each term.
(B) Collect all terms with dy/dx on one side of the equation.
(C) Factor out dy/dx as a common factor.
(D) Move all remaining factors to the other side of the equation using division.

1. Show that if $x^2 + y^2 = 2y^3$, then $\dfrac{dy}{dx} = \dfrac{2x}{6y^2 - 2y}$.

2. Given: $xy^2 + 2y^4 = x^2y$.

(A) Verify that the point $(2, 1)$ is on the curve.

(B) Find the slope of the line tangent to the curve at $(2, 1)$.

Additional Practice

1. Which is the slope of the line tangent to $y^2 + xy - x^2 = 11$ at $(2, 3)$?

(A) $-\dfrac{5}{2}$ (B) 0 (C) $\dfrac{1}{8}$

(D) $\dfrac{4}{7}$ (E) $\dfrac{9}{7}$

2. If $xy^2 - y^3 = x^2 - 5$, then $\dfrac{dy}{dx} =$

(A) $\dfrac{y^2 - 2x}{3y^2 - 2xy}$.   (B) $\dfrac{y^2 - 2x + 5}{3y^2 - 2xy}$.   (C) $\dfrac{2x - 5}{2y - 3y^2}$.

(D) $\dfrac{2x}{2y - 3y^2}$.   (E) $\dfrac{x + y^2}{xy}$.

3. Consider the curve given by $x^2 - x^2y = y^2 - 1$.

   (A) Show that $\dfrac{dy}{dx} = \dfrac{2x - 2xy}{x^2 + 2y}$.

   (B) Find all points on the curve where $x = 2$. Show there is a horizontal tangent to the curve at one of those points.

4. Given $\ln(1 + y) = \dfrac{1}{2}x^2 + 5$, find $\dfrac{d^2y}{dx^2}$.

***Need More Help With* . . .**   *See* . . .

     Implicit differentiation?   *Calculus*, Section 3.7

Derivative as a Rate of Change

<table>
<tr><td>**AP* Objective:**</td><td>Applications of derivatives

• Interpretation of the derivative as a rate of change in varied applied contexts, including velocity, speed, and acceleration</td></tr>
</table>

Big Picture

Although there are countless applications using the derivative to measure change, one of the most important is its use in relating position, velocity, and acceleration. You need to have a clear understanding of velocity as the derivative of position and acceleration as the derivative of velocity. The AP* test may evaluate this understanding in all three common forms—analytically, graphically, and numerically. You can also count on being tested on the subtle difference between average and instantaneous rates of change. Remember, an average rate of change occurs over an interval, while a derivative (instantaneous) occurs at a single point.

Content and Practice

Recall that the *average rate of change* of the position function is the slope of a secant line on the graph of the position function. It is average velocity and requires no calculus to compute. As the time interval becomes shorter and shorter, the limit of the slope of the secant becomes the slope of the tangent line. This is the instantaneous rate of change of position, *velocity*. Velocity therefore is the derivative of the position function. Velocity can be positive or negative, so we say it has both magnitude and direction. For instance, often a falling object is said to have negative velocity. Similarly, the derivative of velocity is acceleration. Remember also that speed is defined as the absolute value of velocity.

1. The position of a particle as a function of time is given by the equation $s(t) = t^2 - 3t$ for $t \geq 0$, where t is in seconds and s is in inches.

   (A) Find the velocity of the particle at the instant $t = 5$ seconds. Include units.

(B) Find the average velocity over the first 4 seconds of motion. (No calculus needed!)

(C) What is the speed of the particle at $t = 1$ second?

2. Consider the following graph of a velocity function. Velocity is in feet per second and time in seconds.

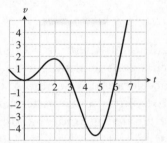

(A) On the domain $0 \leq t \leq 6$ seconds, approximate the time interval(s) when acceleration is positive. Explain your answer. (*Think:* What characteristic of a velocity graph reveals acceleration?)

(B) On the domain $0 \leq t \leq 6$ seconds, approximate the time interval(s) when the speed is decreasing. Explain your answer. (*Hint:* You may want to sketch the absolute value of velocity on the same axes.)

(C) Find an approximation of the average acceleration on the interval $[1, 5]$ seconds. Include units. (*Hint:* No calculus necessary.)

1. The distance of a particle from its initial position is given by
    $s(t) = t - 5 + \dfrac{9}{(t+1)}$, where s is feet and t is minutes. Find the
    velocity at $t = 1$ minute in feet per minute.

    (A) $-\dfrac{5}{4}$

    (B) $\dfrac{13}{4}$

    (C) $\dfrac{1}{2}$

    (D) $-\dfrac{9}{4}$

    (E) $-\dfrac{7}{4}$

2. The distance of a particle from its initial position is measured every 5
    seconds and provided in the table below. Use the data to answer the
    questions that follow.

Time (sec)	0	5	10	15	20
Distance (ft)	0	7	17	25	30

    (A) Estimate the velocity of the particle at $t = 15$ seconds. Include
    units.

    (B) What is the average velocity on the time interval $[5, 20]$ seconds?

3. The graph of the velocity of a particle is given below. On the same
    axes, draw the acceleration graph.

    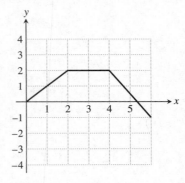

4. The number of liters of water remaining in a tank t minutes after the tank has started to drain is $R(t) = 2t^3 - 20t^2 - 72t + 820$. At what moment is the water draining the fastest?

(A) 0 min   (B) 2 min   (C) $3\frac{1}{3}$ min

(D) $5\frac{1}{3}$ min   (E) It drains at the same rate the whole time.

Need More Help With . . .	See . . .
Derivative as a rate of change?	*Calculus*, Section 3.4

Slope Fields

AP* Objective:

Applications of derivatives
- Geometric interpretations of differential equations via slope fields and the relationships between slope fields and solution curves for differential equations

Big Picture

As you have learned, the solution to a differential equation of the form $dy/dx = f(x)$ is found by integration. When the integral is indefinite, your teacher has most likely emphasized the "+C" at the end of your answer. This is because there are an infinite number of solution curves, all differing by a simple constant. If you just slide a graph up or down, it does not change the graph's slope at any given x-value. Therefore, it is true for all those solutions that at any given value of x, their derivatives (and thus their slopes) are the same. A slope field is a series of very small segments representing the slopes of those solution curves at various points throughout the coordinate plane. This allows us to see the family of solution curves even when we cannot calculate an antiderivative. You should be able to plot a slope field given a differential equation, match a slope field to a differential equation, and draw a particular solution onto a slope field given an initial condition.

Content and Practice

Consider the differential equation $dy/dx = \frac{1}{2}x$. To create a slope field, we calculate the slope at various points in the coordinate plane and plot a small segment with that slope at the given location.

x	y	$\frac{dy}{dx} = \frac{1}{2}x$
-1	0	-0.5
-1	1	-0.5
-1	2	-0.5
0	0	0
0	1	0
0	2	0
1	0	0.5
1	1	0.5
1	2	0.5
2	0	1
2	1	1
2	2	1

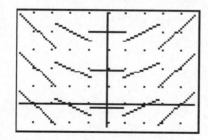

Grid marks are every 0.5 units

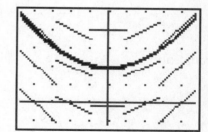

One particular solution plotted on the field,
$y = \frac{1}{4}x^2 + 1$.

Notice that the slopes at various points are dependent only on the value of x. As we look at a column of marks (y-changing), the slopes do not change. This is not always the case, but in this first example, only x

appears in the differential equation. What you should notice in the plot is that the pattern of slope marks begin to look like a series of parabolas. This is true because the solution to the given differential equation is $y = \frac{1}{4}x^2 + C$. You are looking at the changing slopes of the family of translated parabolas. In the second figure, an equation with $C = 1$ has been plotted on the slope field.

1. Given $dy/dx = 1/x$, create a slope field at the 20 given points on the grid.

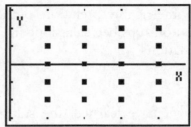

The grid marks are every 1 unit.

2. Indicate which differential equation is represented in the slope field graph. *Briefly* explain your choice.

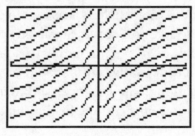

$x: [-6, 6]$ $y: [-4, 4]$

(A) $\dfrac{dy}{dx} = x^3$ (B) $\dfrac{dy}{dx} = \sqrt[3]{x}$ (C) $\dfrac{dy}{dx} = \tan^{-1} x$

(D) $\dfrac{dy}{dx} = x^{-2/3}$ (E) $\dfrac{dy}{dx} = x^{2/3}$

Explanation:

3. A differential equation may also be a function of just y or a combination of x and y. As an example, the slope field below was created from the differential equation $dy/dx = x/y$. Describe any patterns you notice for regions of positive, negative, or zero slope as they may be determined by values of x and y.

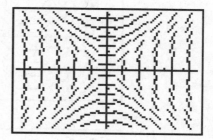

4. Create a slope field on the 12 points in the graph for the differential equation $dy/dx = y^2(x - 1)$.

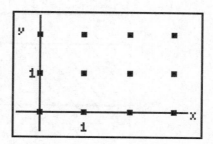

Additional Practice

1. On each slope field, draw the solution curve that satisfies the initial condition $f(-1) = 2$. (Marks on the axes are every 1 unit.)

(A) 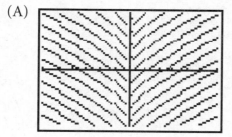   (B)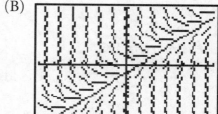

2. Which of the following slope fields could be a solution to the differential equation $dy/dx = x^{1/3}$? Briefly explain your choice. (All windows are $[-4.7, 4.7]$ and $[-3.1, 3.1]$.)

(A)

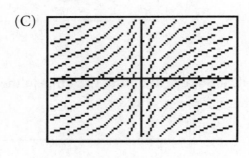

(B)

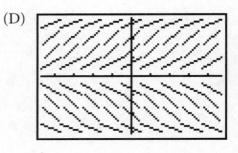

(C)

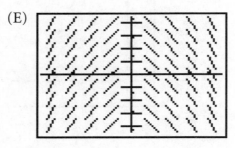

(D)

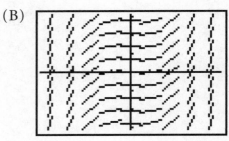

(E)

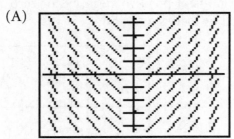

Explanation:

Need More Help With . . .

Slope fields?

See . . .

Calculus, Section 6.1

2004 AP* Calculus AB Exam, Problem 6

Euler's Method

AP* Objective:

Applications of derivatives
• Numerical solutions to differential equations using Euler's method

Big Picture

Euler's method is another strictly AP* Calculus BC topic. Like a tangent line, Euler's method is used to numerically approximate a function near a known point but is generally more accurate over a wider domain. An improved Euler's method is presented in the *Calculus* text but is not part of the AP* Calculus curriculum.

Content and Practice

Early in the course you learned that a tangent line will approximate a function over varying intervals depending on the shape of the function, but often that tangent line became a poor approximation of the function as the slope of the function changed while the slope of the tangent did not. Euler's method basically adjusts the slope of the tangent to more closely follow the curve of the function. Imagine a line tangent to a function. To approximate the function, you follow the line over a designated short domain, called Δx. At that point, you "change course" and follow a path parallel to the slope of the function based on your new location. It is useful to organize your information in a table, as shown.

(x, y)	$f'(x, y)$	Δx or h	$\Delta y = f'(x, y)\Delta x$	$(x + \Delta x, y + \Delta y)$

Graphically, connecting consecutive Euler points produces a segmented curve, which will follow the shape of the curve with varying degrees of accuracy over a limited domain.

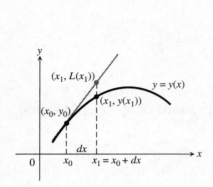

The first Euler step approximates $y(x_1)$ with $L(x_1)$.

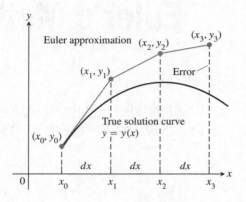

Three steps in the Euler approximation to the solution of the initial value problem $y' = f(x, y)$, $y(x_0) = y_0$. As we take more steps, the errors involved usually accumulate, but not in the exaggerated way shown here.

1. Given that $f(1) = 1$, and $dy/dx = 2x + 1$, generate the next two points of the Euler line using $\Delta x = 0.5$. Show your work in the table provided. If your teacher has given you a program to generate Euler points on your calculator, use it to verify your results. This happens to be a differential equation you can solve analytically, so find the specific function and sketch the function and the Euler line together on the given coordinate plane.

(x, y)	$f'(x, y)$	Δx or h	$\Delta y = f'(x, y)\Delta x$	$(x + \Delta x, y + \Delta y)$
$(1, 1)$		0.5		
		0.5		

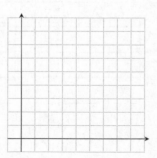

2. In the previous problem, what could be changed in the process of generating points to increase the accuracy of the Euler line?

Often, the differential equation will be a function of both x and y, and you may not be able to solve it by simple separation of variables. Euler's method is particularly useful in these cases to help visualize the function around the known point.

3. Given $f(0) = 1$ and $dy/dx = 2x + y$, use $\Delta x = 0.5$ to generate the next three Euler points. Sketch the curve on the given coordinate plane. Then plot the solution curve $y = 3e^x - 2(x + 1)$ on the same graph.

(x, y)	$f'(x, y)$	Δx or h	$\Delta y = f'(x, y)\Delta x$	$(x + \Delta x, y + \Delta y)$
		0.5		

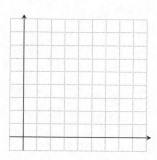

Additional Practice

1. Given $f'(x, y) = \frac{x}{y}$, $f(2) = 1$, and using $\Delta x = 0.5$, which of the following is the Euler approximation of $f(3)$?
    (A) 2.000
    (B) 2.449
    (C) 2.625
    (D) 2.970
    (E) 3.000

2. The error in generating Euler approximations is the absolute value of the difference between the Euler value and the actual function value at a given x. If $dy/dx = 2x + 3$ and $(1, 3)$ lies on the curve, use $\Delta x = 0.2$ and find the error when approximating $f(1.4)$.

(A) 0.04   (B) 0.08   (C) 1.08

(D) 1.40   (E) 3.40

Need More Help With . . .   *See* . . .

Euler's method?   *Calculus,* Section 6.1

L'Hôpital's Rule

AP* Objective:

Applications of derivatives
• L'Hôpital's Rule, including its use in determining limits and convergence of improper integrals and series

Big Picture

There are times when we come across limits that seem to have no obvious solution—for example, a fraction whose numerator and denominator are both approaching zero. A form such as this may not lend itself to easy evaluation. *L'Hôpital's Rule* can often help evaluate these difficult limits. This is a topic required only in the AP* Calculus BC curriculum.

Content and Practice

When taking limits, there is a group of forms that sometimes require the use of L'Hôpital's Rule. These are called *indeterminate forms*. We look for limits whose terms approach any of the following forms: $0/0$, ∞/∞, $\infty \cdot 0$, $\infty - \infty$, 1^{∞}, 0^0, or ∞^0. Students often look at a form like $\infty \cdot 0$ and conclude that it must be zero because "zero times any number is zero." It is vital to remember when you see these forms that they are *limits* and their terms are *approaching* the values, not actually exactly equal to them.

A simple example of this is the limit $\lim\limits_{x \to \infty} \left(x \cdot \dfrac{2}{x} \right)$ The first factor is getting infinitely large while the second factor is heading toward zero, thus we have the form $\infty \cdot 0$, but the limit is not 0. In fact, some small changes can produce a wide variety of results. A simple reduction of the product makes it clear that the limit is 2. If the numerator was any other number, we would get a different limit value. If the denominator was x^2, the limit would be zero. If the first factor was x^2, the limit would grow infinitely large.

L'Hôpital's rule is stated as follows.

> ## L'Hôpital's Rule
>
> Suppose that $f(a) = g(a) = 0$, that f and g are differentiable on an open interval I containing a, and that $g'(x) \neq 0$ on I if $x \neq a$. Then
>
> $$\lim_{x \to a} \frac{f(x)}{g(x)} = \lim_{x \to a} \frac{f'(x)}{g'(x)}.$$

If a limit has any one of the first four indeterminate forms listed above, L'Hôpital's Rule can be applied. The forms $\infty \cdot 0$ and $\infty - \infty$ must be rewritten as a fraction. The remaining three forms 1^∞, 0^0, ∞^0 require finding the limit of the natural logarithm of the expression and using that result as an exponent on e, as required in problems 5, 6, and 7 below.

Find the following limits.

1. $\displaystyle \lim_{x \to \infty} \frac{\ln(x+1)}{\ln(3+x^2)} =$

2. $\displaystyle \lim_{x \to 0} \frac{\sin x - 2x}{e^{3x} - 1} =$

3. $\displaystyle \lim_{x \to 0^+} (\csc x - \cot x) =$

4. $\displaystyle \lim_{x \to \infty} \left(\frac{1}{x} \right) [\ln(x+1) + 5x] =$

For the next problems, keep in mind that if $\displaystyle \lim_{x \to a} [\ln f(x)] = L$ then $\displaystyle \lim_{x \to a} f(x) = e^{\ln f(x)} = e^L$.

5. $\displaystyle \lim_{x \to 3^+} (x-3)^{1/(x-3)} =$

6. $\displaystyle \lim_{x \to 0^+} \sin(x)^x =$

7. $\displaystyle \lim_{x \to \infty} \left(1 + \frac{2}{3x} \right)^x =$

1. $\lim\limits_{x\to 0}\dfrac{\sqrt[3]{1+x}-\dfrac{x}{3}-1}{x^2}=$

    (A) -1 (B) $-\dfrac{1}{9}$ (C) 0

    (D) $\dfrac{1}{3}$ (E) $\dfrac{1}{2}$

2. $\lim\limits_{x\to\infty}\dfrac{\ln\sqrt{x}}{\ln(2+3x)}=$

    (A) 0 (B) $\dfrac{1}{3}$ (C) $\dfrac{1}{2}$

    (D) $\dfrac{2}{3}$ (E) 1

3. $\lim\limits_{x\to 0}(e^x+2x)^{1/x}=$

    (A) 0 (B) 1 (C) e

    (D) e^2 (E) e^3

Need More Help With . . .	*See . . .*
L'Hôpital's Rule?	*Calculus*, Section 8.2

Basic Derivatives

Computation of derivatives

• Basic rules for the derivative of sums, products, and quotients of functions

Big Picture

Basic derivative formulas are the building blocks of the mechanics of calculus. Without the ability to differentiate functions correctly, you will be at a road-block in the course. The basic formulas for derivatives must be committed to memory. These formulas will be tested in the Multiple Choice portion of the AP* exam and will also be necessary to complete certain portions of the Free Response portion of the exam. Additionally, success in integral calculus is strongly dependent on knowing these formulas.

Content and Practice

As a means to review some of your basic derivative formulas, complete the following table by filling in each empty cell. If you have not yet studied all these formulas, it is still productive to look them up and begin committing them to memory. As you move through Chapter 3 of the *Calculus* text, the number of formulas will increase and will address the use of the Chain Rule when differentiating.

	Function	Derivative
1.	$y = x^n$ (n is any real number.)	$\dfrac{dy}{dx} = nx^{n-1}$
2.	$y = k$ (k is a constant.)	
3.	$y = \sin x$	
4.	$y = \cos x$	
5.	$y = \tan x$	
6.		$\dfrac{dy}{dx} = -\csc^2 x$

continued on next page

	Function	Derivative
7.	$y = \sec x$	
8.		$\frac{dy}{dx} = -\csc x \cot x$
9.	$y = \ln x$	
10.		$\frac{dy}{dx} = e^x$

Not only should the formulas be known and easily applied, the concept of a derivative as slope of a curve and as an instantaneous rate of change must also be understood. Any time an independent and dependent variable exist, an instantaneous rate of change of one variable with respect to the other can be examined. The most common independent variable is x, but it can be any variable we choose. We just differentiate with respect to that variable. For example, the area of a circle is dependent on the radius. With $A = \pi r^2$, we can express the instantaneous rate of change of the area with respect to the radius. Differentiating the area equation produces $dA/dr = 2\pi r$.

Practice taking a few derivatives other than differentiating y with respect to x.

11. $P = \tan w$ $\qquad \frac{dP}{dw} =$

12. $V = \frac{1}{3}\pi r^2 h$ (Consider h constant.) $\qquad \frac{dV}{dr} =$

13. $E = mc^2$ (c is constant.) $\qquad \frac{dE}{dm} =$

14. $S = 6t^2$ $\qquad \frac{ds}{dt} =$

Additional Practice

1. What is the slope of the graph of $y = \sin x$ at $x = \pi/3$?

(A) $-\dfrac{\sqrt{3}}{2}$ (B) $-\dfrac{1}{2}$ (C) 0

(D) $\dfrac{1}{2}$ (E) $\dfrac{\sqrt{3}}{2}$

2. What is the instantaneous rate of change of y with respect to x on the graph of $y = e^x$ at $x = a$?

(A) 0   (B) e (C) a

(D) e^a (E) e^x

3. The area of a circle is $A = \pi r^2$. How does the instantaneous rate of change of the area with respect to the radius when $r = 2$ compare to the average rate of change of the area as the radius changes from $r = 1$ to $r = 3$?

(A) The instantaneous rate of change is twice the average rate of change.

(B) The instantaneous rate of change is equal to the average rate of change.

(C) The instantaneous rate of change is half the average rate of change.

(D) The instantaneous rate of change is three times the average rate of change.

(E) Their relationship cannot be determined from the given information.

4. Find the equation of the line tangent to the graph of $y = \sqrt{x}$ at the point on the curve where the y-coordinate is exactly one-third the value of the x-coordinate given $x > 0$. Show the work that leads to your answer.

Need More Help With . . .   *See . . .*

Derivative formulas?   *Calculus,* Sections 3.3 and 3.5

Memorizing?   Make a set of flash cards to study.

Understanding the concept of a derivative?   Ask your teacher for an explanation of the geometric and analytic difference between average rate of change and instantaneous rate of change.

Derivative Rules

AP* Objective:
Computation of derivatives
• Basic rules for the derivative of sums, products, and quotients of functions

Big Picture

Frequently, the functions we need to differentiate are irreducible sums, differences, products, or quotients of functions. Mastery of the formulas for these cases is another key skill in the early part of any calculus course. These formulas will be tested in the Multiple Choice portion of the AP* Calculus exam and will also be necessary to complete certain portions of the Free Response portion of the exam. Additionally, success in integral calculus is strongly dependent on knowing these formulas.

Content and Practice

As a means to review derivative formulas, complete the following table below by filling in each empty cell. If you have not yet studied all these formulas, it is still productive to look them up and begin committing them to memory. As you move through Chapter 3 of the *Calculus* text, the number of formulas will increase and will address the use of the Chain Rule when differentiating.

Formula Name	Derivative Formula	Word Description
Difference Formula	$y = f(x) - g(x)$ $\dfrac{dy}{dx} = f'(x) - g'(x)$	The derivative of a difference of functions is the difference of their individual derivatives.
Sum Formula	$y = f(x) + g(x)$ $\dfrac{dy}{dx} = f'(x) + g'(x)$	
Product Rule	$y = f(x) \cdot g(x)$ $\dfrac{dy}{dx} = f'(x) \cdot g(x) + f(x) \cdot g'(x)$	
Quotient Rule	$y = \dfrac{f(x)}{g(x)}$ $\dfrac{dy}{dx} = \dfrac{g(x) \cdot f'(x) - f(x) \cdot g'(x)}{[g(x)]^2}$	

Complete the following derivatives using the appropriate formula. The most common independent variable is x, but it can be any variable we choose. We just differentiate with respect to that variable.

1. $y = 5x^3 + Ln(x)$ $\dfrac{dy}{dx} =$

2. $y = x^3 \cdot Ln(x)$ $\dfrac{dy}{dx} =$

3. $s = \dfrac{5}{t^4}$ $\dfrac{ds}{dt} =$

4. $k = \dfrac{\sin p}{\sqrt{p}}$ $\dfrac{dk}{dp} =$

5. $w = z^4 - (z + 2)^7 \cdot \cos z$ $\dfrac{dw}{dz} =$

6. $y = x^2(x^3 - 2)$ Find $\dfrac{dy}{dx}$ two different ways. Distribute x^2 before differentiating. Use product rule on the two factors. Show the two answers are equivalent.

Additional Practice

1.   Given $f(x) = e^x \cdot \cos x$, choose a solution to $f'(x) = 0$:

    (A) 0   (B) $\dfrac{\pi}{4}$ (C) $\dfrac{\pi}{2}$

    (D) $\dfrac{3\pi}{4}$ (E) π

2. Find the equation of the line tangent to $y = \dfrac{x + 3}{x^2 + 1}$ at $x = 1$.

(A) $y = -\dfrac{5}{2}x + \dfrac{9}{2}$ (B) $y = \dfrac{5}{2}x - \dfrac{1}{2}$

(C) $y = \dfrac{1}{2}x + \dfrac{3}{2}$ (D) $y = -\dfrac{3}{2}x + \dfrac{1}{2}$

(E) $y = -\dfrac{3}{2}x + \dfrac{7}{2}$

Use the table below for solving numbers 3 & 4.

x	$f(x)$	$g(x)$	$f'(x)$	$g'(x)$
1	4	2	5	$\dfrac{1}{2}$
3	7	-4	$\dfrac{3}{2}$	-1

3. The value of $\dfrac{d}{dx}(f \cdot g)$ at $x = 3$ is:

(A) $\dfrac{5}{2}$ (B) $-\dfrac{3}{2}$ (C) -13

(D) 12   (E) $\dfrac{21}{2}$

4. The value of $\dfrac{d}{dx}\left(\dfrac{f}{g}\right)$ at $x = 1$ is:

(A) 2   (B) 3   (C) 5

(D) 6   (E) 10

Need More Help With . . .

Formulas used here?
Practice problems?

See . . .

Section 3.3 of your *Calculus* text. Make up some of your own product and quotient rule problems and check your answers with a computer algebra system such as that found on a TI-89.

Chain Rule

AP* Objective:

Computation of derivatives
• Chain Rule and implicit differentiation

Big Picture

In precalculus, you practiced decomposing functions. This skill is useful in understanding and applying the Chain Rule. Using the Chain Rule correctly is one of the most important skills in the calculus course. It is applied to derivatives and integrals throughout much of the rest of the year.

Content and Practice

Suppose we wanted to find dy/dx if $y = \sin(x^2 + 3x)$. Think of y as $f(x)$ and recognize $f(x)$ as a composite of two functions: $f(x) = h(g(x))$, where $g(x) = x^2 + 3x$ and $h(x) = \sin x$. In this case we can write $y = \sin u$, where $u = x^2 + 3x$. Using the Chain Rule, we have

$$\frac{dy}{dx} = \frac{dy}{du} \cdot \frac{du}{dx}$$

$$= \frac{d(\sin u)}{du} \cdot \frac{d(x^2 + 3x)}{dx}$$

$$= \cos u \cdot (2x + 3)$$

$$= \cos(x^2 + 3x) \cdot (2x + 3).$$

The *Calculus* text also presents this as the derivative of the "outside" function, sine, times the derivative of the "inside" function, the polynomial. In some cases, it is also possible to have more than just two factors involved in the Chain Rule, for example, $dy/dx = dy/du \cdot du/dt \cdot dt/dx$.

1. Find dy/dx if $y = (4x + 2)^3$, using y as a composite of $y = u^3$ and $u = 4x + 2$.

2. Following the first example, write $h = \sqrt{1 - \sin v}$ as a composition of two functions, g (the inner function of v) and h (the outer function of g), so that $h(v) = h(g(v))$. Use your answers to find dh/dv using $dh/dv = dh/dg \cdot dg/dv$.

3. Use composition to show $dy/dt = 6t \sin^2 (t^2 + 5) \cos (t^2 + 5)$ if $y = [\sin (t^2 + 5)]^3$. This is an example where you may have three factors involved in the process.

4. What is $f'(x)$ if $f(x) = \sin^2 (3x)$?
   (A) $2 \sin 3x$
   (B) $6 \sin 3x$
   (C) $2 \sin (3x) \cdot \cos (3x)$
   (D) $6 \sin (3x) \cdot \cos (3x)$
   (E) $2 \sin 3x + 3 \cos 3x$

5. The derivative formulas you learned recently were simple functions of x and did not require the Chain Rule. Those functions and numerous additional functions now take on the form seen in the table. A few lines are completed for you. Fill in the remaining blanks, then check the formulas in your book. If there are formulas you have not yet learned, look them up and begin committing them to memory.

	Function	Derivative
	$y = u^n$ (n is any real number.)	$\dfrac{dy}{dx} = nu^{n-1}\dfrac{du}{dx}$
(a)	$y = k$ (k is a constant.)	
	$y = \sin u$	$\dfrac{dy}{dx} = \cos u \cdot \dfrac{du}{dx}$
(b)	$y = \cos u$	
(c)	$y = \tan u$	
(d)		$\dfrac{dy}{dx} = -\csc^2 u \cdot \dfrac{du}{dx}$
(e)	$y = \sec u$	
(f)		$\dfrac{dy}{dx} = -\csc u \cot u \cdot \dfrac{du}{dx}$
(g)	$y = e^u$	
(h)		$\dfrac{dy}{dx} = \dfrac{1}{u} \cdot \dfrac{du}{dx}$
(i)	$y = a^u$ ($a > 0, a \neq 1$)	
(j)	$y = \sin^{-1} u$ (This is inverse sine.)	
(k)		$\dfrac{dy}{dx} = \dfrac{-1}{1 + u^2} \cdot \dfrac{du}{dx}$
(l)	$y = \sec^{-1} u$	
(m)		$\dfrac{dy}{dx} = \dfrac{-1}{|u|\sqrt{u^2 - 1}} \cdot \dfrac{du}{dx}$
(n)	$y = \tan^{-1} u$	
(o)	$y = \cos^{-1} u$	
(p)	$y = \log_a u$	

Additional Practice

The functions f and g and their derivatives have the following values at $x = 1$ and $x = 2$.

x	$f(x)$	$g(x)$	$f'(x)$	$g'(x)$
1	3	2	0	$\frac{3}{4}$
2	7	-4	$\frac{1}{3}$	-1

Use the information above to find the value of the first derivative of the following at the given value of x.

1. $[f(x)]^2$ at $x = 2$

    (A) 0   (B) $\frac{2}{3}$ (C) $\frac{14}{3}$

    (D) 6   (E) None of these

2. $f(g(x))$ at $x = 1$

    (A) $-\frac{1}{3}$ (B) 0   (C) $\frac{1}{4}$

    (D) $\frac{9}{4}$ (E) $\frac{21}{4}$

Need More Help With . . .

Decomposing functions?

Basic derivatives?

Chain Rule?

See . . .

Precalculus, Section 1.4

Calculus, Sections 3.3, 3.5, 3.8, and 3.9

Calculus, Section 3.6

Derivatives of Parametric, Polar, and Vector Functions

AP* Objective:

Computations of derivatives

• Derivatives of parametric, polar, and vector functions

Big Picture

The rules for derivatives apply to parametric, polar, and vector functions. In addition, you should also know the Chain Rule for parametric equations.

Chain Rule for Parametric Equations

If x and y are given as parametric functions of t, then

$$\frac{dy}{dx} = \frac{dy/dt}{dx/dt} \text{ and } \frac{d^2y}{dx^2} = \frac{d/dt(dy/dx)}{dx/dt}$$

To differentiate a vector function, simply differentiate the components.

$$\text{If} \quad \mathbf{r} = x(t)\mathbf{i} + y(t)\mathbf{j}, \quad \text{then} \quad \frac{d\mathbf{r}}{dt} = \frac{dx}{dt}\mathbf{i} + \frac{dy}{dt}\mathbf{j}.$$

Content and Practice

1. If $x(t) = t^2 - t$ and $y(t) = t^3 + 1$, find dy/dx and d^2y/dx^2 in terms of t.

$$\frac{dy}{dx} = \frac{3t^2}{2t - 1}$$

$$\frac{d^2y}{dx^2} = \frac{\dfrac{d}{dt}\left(\dfrac{3t^2}{2t-1}\right)}{2t-1} = \frac{\dfrac{6t(t-1)}{(2t-1)^2}}{(2t-1)} = \frac{6t(t-1)}{(2t-1)^3}$$

2. If $x(t) = 4 \cos t$ and $y(t) = \sin 2t$, find d^2y/dx^2 when $t = \pi/2$.

3. If f is a vector-valued function defined by $f(t) = (e^{2t}, \cos t)$, then $f''(t) = \underline{\hspace{1cm}}$.

Additional Practice

1. An object moving along a curve in the xy-plane has position $(x(t), y(t))$ at time $t \geq 0$ with $dx/dt = 3 + \sin t^2$. The derivative dy/dt is not explicitly given. At time $t = 2$, the object has position $(1, 3)$ and $dy/dt = -5$. Find an equation for the line tangent to the curve at the point $(x(2), y(2))$.

2. If $x(t) = t^2 - t$ and $y(t) = \sqrt{3t + 1}$, then dy/dx at $t = 1$ is

   (A) $-\dfrac{1}{2}$.   (B) $\dfrac{1}{2}$.   (C) $\dfrac{3}{4}$.

   (D) 1.   (E) 2.

3. If $\mathbf{f}$ is a vector-valued function defined by $\mathbf{f}(t) = \ln(t)\mathbf{i} + \sqrt{t}\,\mathbf{j}$, then $\mathbf{f}''(1) =$

   (A) $-\mathbf{i} - \dfrac{1}{4}\mathbf{j}$.   (B) $\mathbf{i} - \dfrac{1}{4}\mathbf{j}$.   (C) $-\mathbf{i} - \dfrac{1}{2}\mathbf{j}$.

   (D) $\mathbf{i} - \dfrac{1}{2}\mathbf{j}$.   (E) $\mathbf{i} + \dfrac{1}{2}\mathbf{j}$.

Need More Help With . . .	*See . . .*
Derivatives of parametric or vector functions?	*Calculus,* Sections 10.1, 10.2

Riemann Sums

AP* Objective: Interpretations and properties of definite integrals

• Computation of Riemann sums using left, right, and midpoint evaluation points

• Definite integral as a limit of Riemann sums over equal subintervals

Big Picture

Many important tasks in calculus involve finding the area under a curve. Areas represent answers to many real-world problems such as the distance traveled by an object with variable velocity or the work done by a variable force. The area under a curve can be divided into narrow strips. Each of these strips can be approximated by a narrow rectangle. The sum of the areas of the rectangles approximates the area under the curve. This sum is called a *Riemann sum*. The exact area can be found by taking a limit of the Riemann sum as the widths of the rectangles approach zero. This limit is called a *definite integral*.

Content and Practice

The area under the curve $y = 4 - x^2$ on the interval $-2 \le x \le 2$ can be approximated by three different rectangular methods. Using ten subdivisions, the graphs below visually suggest these rectangular approximation methods.

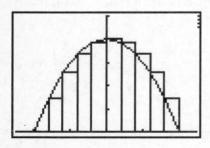

Left-hand rectangles

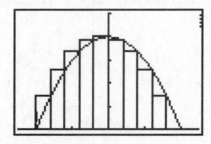

Right-hand rectangles

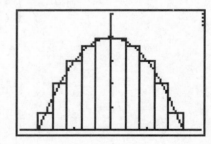

Midpoint rectangles

1.   Approximate the area bounded by $f(x) = 4 - x^2$ and the *x*-axis using

(A) left-hand rectangles and four equal subdivisions.

(B)   right-hand rectangles and four equal subdivisions.

(C)   midpoint rectangles and four equal subdivisions.

2.   The table shows the velocity of a remote-controlled toy car as it traveled down a hallway for 10 seconds.

Time (sec)	0	1	2	3	4	5	6	7	8	9	10
Velocity (in./sec)	0	6	10	16	14	12	18	22	12	4	2

Estimate the distanced traveled by the car using 10 subintervals of length 1 and the methods shown.

(A)   Left-hand rectangles

(B)   Right-hand rectangles

Additional Practice

1.   The table shows the rate in liters/min at which water leaked out of a container.

Time (min)	0	1.2	2.3	3.8	5.4
Rate (liters/min)	5.6	4.3	3.1	2.2	1.5

A right-hand Riemann sum is computed using the four subintervals indicated by the data in the table. This Riemann sum estimates the total amount of water that has leaked out of the container. What is the estimate?

(A)   12.70 liters   (B)   14.27 liters   (C)   16.70 liters

(D)   16.95 liters   (E)   19.62 liters

2. The temperature, in degrees Celsius (°C), of a turkey in an oven is a continuous function of time t. Some values of this function are given in the table.

Time (min)	0	5	10	15	20	25
Temperature (°C)	24	76	106	124	135	141

Approximate the average temperature (in degrees Celsius) of the turkey over the time interval $0 \le t \le 25$ using a left-hand Riemann sum with subintervals of length 5 minutes.

Need More Help With . . .	*See . . .*

Riemann sums?

There is also a calculator program to compute Riemann sums in the *Technology Resource Manual* that accompanies *Calculus*. This program is called RAM. The average value of a function is defined in *Calculus* Section 5.3.

Calculus, Sections 5.1 and 5.2

Definite Integral of a Rate of Change

AP* Objective:

Interpretations and properties of definite integrals

• Definite integral of the rate of change of a quantity over an interval interpreted as the change of the quantity over the interval:

$$\int_a^b f'(x)\, dx = f(b) - f(a)$$

Big Picture

If you are given a function that describes the rate of change of a quantity, the definite integral of this function gives the net change in the quantity. The definite integral of a rate of change can be used to find the net change in quantities such as distance, velocity, growth, decay, production, and consumption.

Content and Practice

Suppose the velocity in cm/sec of a particle moving along the horizontal x-axis is given by the equation $dx/dt = f'(t) = 4 - t^2$ for $0 \le t \le 3$. The figure below shows the graph of this velocity.

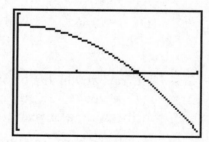

The velocity is the rate of change of position. Integrating this rate of change will give the net change in position or, in other words, the net change in distance from the starting point. The net change in distance for this particle after 3 seconds is

$$\int_0^3 (4 - t^2)\, dt = 3 \text{ cm.}$$

1. If the starting point for the particle is $x = 1$, find the position of the particle after

   (A) 3 seconds

   (B) 2 seconds

2. From 1990 to 2004 the production of apples in a certain orchard was $P(t) = 1.3 + 1.025^t$ thousand bushels per year, where t is the number of years from the beginning of 1990. Find the total number of bushels produced from the beginning of 1990 to the beginning of 2004.

Additional Practice

1. Water is pumped from a storage tank to meet the needs of a town. The flow of water from the tank is given by $C(t) = 25e^{-0.05(t-15)^2}$ thousand gallons per hour, where t is the number of hours since midnight. Which of the following best approximates the total water consumption for one day (in thousands of gallons)?
   (A) 33.964 (B) 164.202 (C) 197.727
   (D) 198.166 (E) 202.144

2. The rate at which a bison herd is increasing is given by $B(t) = 26.7 + 1.036^t$ bison per year where t is the time in years since the beginning of year 2000.

   (A) If the herd contains 756 bison at the beginning of 2000, predict the size of the herd at the beginning of 2015.

(B) Predict the average annual increase in the bison herd from 2000 to 2015.

Need More Help With . . .   *See . . .*

Integrating a rate of change?   *Calculus,* Section 7.1

You should pay particular
attention to the difference between
displacement and *total distance.*
Question 2 in the Additional Practice
section involved finding the average
value of a function, which is defined
in *Calculus,* Section 5.3.

Basic Properties of Definite Integrals

AP Objective: Interpretations and properties of definite integrals
• Basic properties of definite integrals (Examples include additivity and linearity.)

Big Picture

The basic properties of definite integrals allow you to evaluate variations on a definite integral such as switching limits of integration, multiplying by a constant, adding two definite integrals, or breaking a definite integral over an interval into separate parts.

Content and Practice

Rules for Definite Integrals

1. $\int_b^a f(x)\, dx = -\int_a^b f(x)\, dx$

2. $\int_a^a f(x)\, dx = 0$

3. $\int_a^b k f(x)\, dx = k \int_a^b f(x)\, dx$

4. $\int_a^b (f(x) \pm g(x))\, dx = \int_a^b f(x)\, dx \pm \int_a^b g(x)\, dx$

5. $\int_a^b f(x)\, dx + \int_b^c f(x)\, dx = \int_a^c f(x)\, dx$

6. $\min f (b - a) \leq \int_a^b f(x)\, dx \leq \max f (b - a)$

7. $f(x) \geq g(x)$ on $[a, b] \Rightarrow \int_a^b f(x)\, dx \geq \int_a^b g(x)\, dx$

8. $f(x) \geq 0$ on $[a, b] \Rightarrow \int_a^b f(x)\, dx \geq 0$

---

1. Let $\int_1^2 f(x)\,dx = -3$, $\int_1^5 f(x)\,dx = 5$, and $\int_1^5 g(x)\,dx = 9$. Find

    (A) $\int_2^2 g(x)\,dx$

    (B) $\int_1^2 (3f(x) + 1)\,dx$

    (C) $\int_2^5 f(x)\,dx$

2. If $f(x)$ is continuous on $[1, 3]$ and $2 \le f(x) \le 4$, what is the greatest possible value of $\int_1^3 f(x)\,dx$?

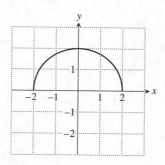

3. The graph of f is the semicircle shown above. Let g be the function given by $\int_0^x f(t)\,dt$. What is the value of $g(-2)$?

Additional Practice

1. If f is increasing on the interval $[a, b]$, which of the following must be true?

    I. $\int_a^b f(x)\,dx \le f(b)(b - a)$

    II. $\int_a^b f(x)\,dx \ge f(a)(b - a)$

    III. $\int_a^b f(x)\,dx \ge 0$

    (A) None (B) I only (C) II only

    (D) I, II, and III (E) I and II only

2. Suppose that h is continuous and that $\int_{-1}^{1} h(r)\,dr = 0$ and $\int_{-1}^{3} h(r)\,dr = 7$. Find

(A) $\int_{1}^{3} h(r)\,dr$

(B) $-\int_{3}^{1} h(r)\,dr$

Need More Help With . . .   *See . . .*

Properties of definite integrals?   *Calculus,* Section 5.3

Analyzing antiderivaties graphically?   *Calculus,* Section 5.2, 5.4

Applications of Integrals

AP* Objective: Applications of integrals

Big Picture

Applications of integrals occur in a wide variety of settings. Regardless of the setting, the emphasis should be on using the integral of a rate of change to give accumulated change or setting up an approximating Riemann sum and representing its limit as a definite integral.

Content and Practice

Some common applications of integrals on the AP* Calculus Exam include area, volume, average value of a function, distance traveled by a particle, and curve length. (Curve length is only found on the BC exam.) In many cases it may be helpful to slice the problem into small sections and set up the appropriate Riemann sum. This sum can then be used to find the corresponding definite integral.

1.   *Area:* Find the area bounded by the curve $y = 4 - x^2$ and the x-axis.

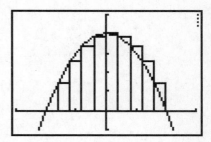

Solution: The area can be approximated by rectangular slices. The Riemann sum for these slices is $\sum_{k=1}^{n}(4 - x^2)\Delta x$. The limit of this sum gives the definite integral $\int_{-2}^{2} (4 - x^2)\, dx = 10\frac{2}{3}$.

2. Find the area bounded by the curves $y = \cos x$, $y = x$, and the y-axis.

3. Find the volume generated when the region bounded by the curves $y = \sqrt{x}$ and $y = \frac{1}{2}x$ is rotated about the x-axis.

4. (BC only) *Curve length:* Set up but do not evaluate the integral that gives the length of the path described by the parametric equations $x = t^2$ and $y = t^3$, where $0 \le t \le 1$.

Additional Practice

1. The base of a solid is a region in the first quadrant bounded by the x-axis, the y-axis, and the line $y = 1 - x$. If cross sections of the solid perpendicular to the x-axis are semicircles, what is the volume of the solid?

2. What is the average value of the function $y = 3x^2\sqrt{x^3 + 1}$ on the interval $[0, 2]$?

   (A) $\dfrac{26}{3}$ (B) $\dfrac{52}{3}$ (C) 18

   (D) 24   (E) $\dfrac{82}{3}$

3. Find the *total distance* traveled by the particle moving along a straight line with velocity $v = \sin \pi t$ for $0 \le t \le 2$.

4.   (BC only) *Polar area:* Find the area inside the polar curve $r = 4\cos\theta$ and outside the polar curve $r = 2$.

5.   (BC only) *Curve length:* Find the length of the curve described by $y = \frac{2}{3}x^{3/2}$ from $x = 0$ to $x = 3$.

(A) $\frac{2}{3}$ (B) $\frac{7}{3}$ (C) $\frac{14}{3}$ (D) $\frac{16}{3}$ (E) 7

Need More Help With . . .   *See . . .*

Average value of a function?   *Calculus,* Section 5.3

Total distance?   *Calculus,* Section 7.1

Area under a curve?   *Calculus,* Section 7.2

Finding volumes?   *Calculus,* Section 7.3

Length of a curve?   *Calculus,* Section 7.4

Length of parametric curves?   *Calculus,* Section 10.1

Polar areas?   *Calculus,* Section 10.3

Fundamental Theorem of Calculus

AP* Objective:

Fundamental Theorem of Calculus
- Use the Fundamental Theorem to evaluate definite integrals
- Use the Fundamental Theorem to represent a particular antiderivative and the analytic and graphical analysis of functions so defined

Big Picture

The Fundamental Theorem of Calculus describes a remarkable connection between integration and differentiation. It is one of the most important ideas in mathematics. The Fundamental Theorem has two parts. The first part says that the definite integral of a continuous function is a differentiable function of its upper limit of integration. It also tells us how to differentiate this definite integral. The second part tells us how to evaluate a definite integral.

Content and Practice

The first part of the Fundamental Theorem says $\frac{d}{dx} \int_a^x f(t)\, dt = f(x)$, where f is a continuous function. The second part says $\int_a^b f(x)\, dx = F(b) - F(a)$, where F is any antiderivative of f.

1. Evaluate $\int_1^e \left(x - \frac{2}{x} \right) dx$.

2. Evaluate $\frac{d}{dx} \int_0^{x^2} \sin t\, dt$.

3. Let $F(x) = \int_0^x f(t)\, dt$, where $f(t)$ is the continuous function whose graph is shown.

(A) Where does F achieve its maximum value? Explain.

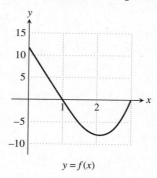

$y = f(x)$

(B) Where does F achieve its minimum value? Explain.

(C) Sketch a possible graph for F on the interval $[0, 3]$.

Additional Practice

1. Evaluate $\dfrac{d}{dx} \int_x^5 2^{t^2}\, dt$.

2. Let f be the continuous function defined on $[-1, 3]$ whose graph, shown below, is comprised of a line segment and semicircle.

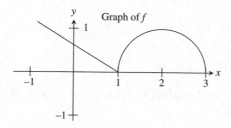

Graph of f

If $g(x) = \int_1^x f(1)\,dt$, use the graph of f on the previous page to

(A)   Find $g(-1)$, $g(1)$, and $g(3)$

(B)   Find intervals on $[-1, 3]$ where g is decreasing.

(C)   Find intervals on $[-1, 3]$ where the graph of g would be concave downward.

3.   Which of the following is the greatest value of x on the interval $[0, 3]$ for which $\int_0^x (t^2 - 3t)\,dt \geq \int_2^x t\,dt$?

(A) 0.56   (B) 0.92   (C) 1.36   (D) 1.57   (E) 1.78

4.   Let f be the differentiable function whose graph is shown. The position at time t (in seconds) of a particle moving along a coordinate axis is $s = \int_0^t f(x)\,dx$ meters.

(A)   What is the particle's position at time $t = 2$?

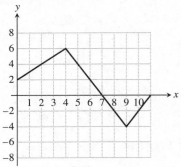

(B)   At what time during the first 11 seconds does the particle's position have its largest value? Justify your answer.

Need More Help With . . .   ***See . . .***

Fundamental Theorem of Calculus?   *Calculus,* Section 5.4

Antiderivative Basics

AP* Objective: Techniques of antidifferentiation
• Antiderivatives following directly from derivatives of basic functions

Big Picture

The second major portion of the calculus course depends largely on the skill of finding antiderivatives. The process can range from simply looking at an expression and knowing the function whose derivative has been taken, to many advanced techniques. You also learn that the limiting process of Riemann sums results in integrals, and the Fundamental Theorem of Calculus connects integrals to antiderivatives.

Content and Practice

The key to successful antidifferentiation is to master the derivatives of basic functions. Below is a list similar to what appeared in the section on basic derivatives, but you will notice three changes: the column headings, the order of the columns, and the presence of a "+C." One challenging aspect of working with antiderivatives is to not confuse which "direction" you are going. Students frequently take a derivative when they think they are taking an antiderivative.

  Fill in the empty spaces in the table. It will help you review the basic relationships and force you to practice going different directions with the process.

		Derivative	Antiderivative
		$n \cdot x^{n-1}$	$x^n + C$ (n is any real number.)
	(A)		$\sin x + C$
	(B)		$-\cos x + C$
	(C)		$\tan x + C$
	(D)	$-\csc^2 x$	
	(E)		$\sec x + C$
	(F)	$-\csc x \cot x$	
	(G)	$\dfrac{1}{x}$	
	(H)		$e^x + C$
	(I)	$a^x \ln a$ (a is a constant.)	

By observation, attempt to determine the antiderivative of each of the following functions. Check your answer by taking its derivative. Remember the Chain Rule when checking your answers. You may need to adjust your answer by multiplying or dividing by a constant.

	Function	Antiderivative
2.	$\sin 2x$	
3.	$8x^3 + \sqrt{x}$	
4.	$3 + e^{5t}$	
5.	$x \cos x^2$	
6.	$8x - \csc x \cot x$	
7.	$\sec^2 5x$	
8.	$6(2x + 7)^5$	
9.	$3^{4x} \ln 3$	

10. In your own words, write a sentence or two expressing what you understand an antiderivative to be.

11. Why is the "+C" placed on antiderivatives when they are not associated with definite integrals?

Additional Practice

1. $\int 2x + 7\, dx =$

   (A) $(x^2 + 7x)\, x + C$ (B) $x^2 + 7x + C$ (C) $x^2 + C$

   (D) $2 + C$ (E) $\dfrac{x^2}{2} + 7x + C$

2. $\int_0^{\pi/6} \sin 2x\, dx =$

   (A) $-\dfrac{1}{2}$ (B) $-\dfrac{1}{4}$ (C) 0

   (D) $\dfrac{1}{4}$ (E) $\dfrac{1}{2}$

3. If $f'(x) = \dfrac{2x}{x^2 + 1}$, then $f(x)$ could be

(A) $3 + \ln(x^2 + 1)$.

(B) $\dfrac{x^2}{\frac{1}{3}x^3 + x} + 8$.

(C) $\dfrac{2 - 2x^2}{(x^2 + 1)^2} + C$.

(D) $x + \ln(x^2 + 1) + C$.

(E) $8 - \ln(1 - x^2)$.

Need More Help With . . .

Definition of an antiderivative?

Working with antiderivatives?

Antiderivative formulas?

See . . .

Precalculus, Section 4.2

Calculus, Sections 5.3 and 5.4

Calculus, Section 6.2 and 6.3

Antidifferentiation by Substitution

AP* Objective:

Techniques of antidifferentiation
- Antiderivatives by substitution of variables (including change of limits for definite integrals), parts, and simple partial fractions (nonrepeating linear factors only)

Big Picture

As integrals become more complicated, it is often not possible to determine an antiderivative by observation. Some work must be done to see the form of the integral. One basic method is substitution of variables. A proper substitution will reveal the form of an integral and make antidifferentiation much easier.

Content and Practice

If an integral is to be handled by substitution, the process requires identifying within the integrand a function and a constant multiple of its derivative. We then substitute a single variable for the main function and the appropriate differential for the remaining factors in the integrand. As stated earlier in this text, the better you know the derivative formulas, the easier the substitution method will be. Study the example below before attempting the subsequent problems.

Suppose we wanted to integrate the indefinite integral

$$\int 4x(x^2 + 5)^8 \, dx.$$

One way would be to expand the power of the binomial, distribute the $4x$, and integrate each of the nine terms individually. In a mere 30 minutes or so the problem would be solved! Substitution provides a more efficient solution.

Solving the Integral by Substitution	
Recognize that $4x$ is a multiple of the derivative of $x^2 + 5$. An exponent on a function is often *not* chosen as part of the main function. We will therefore choose $x^2 + 5$ to be our main function and $4x\,dx$ to be a multiple of its derivative.	$\int 4x(x^2 + 5)^8\,dx$
Notice the integrand contains $4x\,dx$, not $2x\,dx$. There are numerous ways to account for this. We will choose to simply multiply both sides of the differential equation by 2.	Let $u = x^2 + 5$, then $du = 2x\,dx$ so $2\,du = 4x\,dx$.
Substituting for the base of the power function and the other factors in the integrand allows us to see the form of the integrand much more clearly.	$\int u^8 2\,du = 2\int u^8\,du$
Find the antiderivative using the Power Rule for integration.	$2\left(\dfrac{u^9}{9}\right) + C$
Re-substitute for u from our original choice, $u = x^2 + 5$. Clearly this is much more efficient than expanding an eighth power of a binomial!	$\dfrac{2}{9}(x^2 + 5)^9 + C$

Identifying u and du in the integrand is one of the most important skills in the substitution method of integration. It will be helpful to practice. For each of the following integrals, choose which part of the integrand will be u and which factors relate to du.

1. $\int \sin x^3\, 3x^2\,dx$ 2. $\int \dfrac{5\,dx}{4x + 3}$

3. $\int 7\tan^5 x \sec^2 x\,dx$ 4. $\int \dfrac{4x\,dx}{\sqrt{x^2 - 4}}$

5. $\int \csc x^2 \cdot \cot x^2 \cdot 8x\,dx$ 6. $\int \dfrac{3^{\ln(x)}}{x}\,dx$

7. Integrate by the substitution method.

$$\int \frac{6x^2 + 10}{x^3 + 5x} \, dx$$

The substitution method can also be used on definite integrals. When doing so, you should also substitute for the limits of integration. The relationship between u and your choice of main function in the integrand determines the new limits. Initial steps of an example are presented below.

$$\int_0^2 (x^3 + 1)^{3/2} \cdot x^2 \, dx$$
Let $u = x^3 + 1$, so $du = 3x^2 \, dx$, and $\frac{1}{3} du = x^2 \, dx$.

Also, if $x = 0$, then $u = 1$; and if $x = 2$, then $u = 9$.

By substitution into the integrand and limits, the integral becomes

$$\frac{1}{3} \int_{u=1}^{u=9} u^{3/2} \, du.$$

We can now finish the problem without ever substituting back in terms of x.

Additional Practice

1. $\int e^{\cos x} \sin x \, dx =$

(A) $-e^{\cos (x)+1} + C$ (B) $e^{\cos x} + C$ (C) $-e^{\cos x} + C$

(D) $e^{\sin x} + C$ (E) $e^{\sin x} \cos (x) + C$

2. Given: $\int_0^{\pi/6} \sin 2x \cos 2x \, dx$. Letting $u = \sin 2x$, the integral becomes

(A) $\int_0^{1/2} u \, du.$ (B) $\frac{1}{2} \int_0^{1/2} u \, du.$ (C) $2 \int_0^{\sqrt{3}/2} u \, du.$

(D) $\int_0^{\sqrt{3}/2} u \, du.$ (E) $\frac{1}{2} \int_0^{\sqrt{3}/2} u \, du.$

3. Evaluate the integral $\int_1^2 \frac{2x^3 + 1}{x^4 + 2x} \, dx.$

(A) $\ln 20 - \ln 3$ (B) $\ln \sqrt{\frac{20}{3}}$ (C) $\frac{1}{2} \ln 2$

(D) $\ln 2$ (E) None of the above

4.   Suppose $\int_2^4 f(2x)\, dx = 10$. Then $\int_4^8 f(u)\, du =$

(A) 4.   (B) 5.   (C) 10.

(D) 20.   (E) Cannot be determined

Need More Help With . . .   **See . . .**

Integration using the substitution   *Calculus*, Section 6.2
method?

Antidifferentiation by Parts

Techniques of antidifferentiation
- Antiderivatives by substitution of variables (including change of limits for definite integrals), parts, and simple partial fractions (nonrepeating linear factors only)

Big Picture

Antidifferention by parts is strictly an AP* Calculus BC topic. This process must be considered when direct substitution does not work. It is frequently tested in the Multiple Choice portion of the AP* exam.

Content and Practice

There are times when integration by substitution does not work because when we choose a function to substitute for, a multiple of its derivative does not exist in the integrand. In other words, the factors in the integrand consist of a function and the derivative of a different function. It is symbolized by an integral of the form $\int u\,dv$, where u is the primary function and dv is the derivative of another function.

There are two keys to using antidifferentiation by parts successfully. First, recognize when no other integration method will apply. Second, choose the correct parts of the integrand as u and dv. An acronym for the priority order for choosing u is LIPET.

L: Logarithms
I: Inverse trigonometric functions
P: Polynomials
E: Exponential functions
T: Trigonometric functions

Once u and dv have been chosen, du and v must be identified. At this point it is important to keep straight whether you are finding a derivative or an antiderivative. Once you have determined all four parts, substitute them into the formula $\int u\,dv = uv - \int v\,du$.

In the following integrals, identify what you would choose as u and dv. Do not attempt to integrate. One of the five integrals can be done by direct substitution. For that one, choose u and du.

1. $\int x^2 \ln x \, dx$

2. $\int x \tan^{-1} x \, dx$

3. $\int x \cos x \, dx$

4. $\int \dfrac{e^{\sqrt{x}} \, dx}{\sqrt{x}}$

5. $\int 5^x x \, dx$

6.   Integrate by parts: $\int x \ln x^2 \, dx$.

As shown in the next examples, when antidifferentiation by parts is necessary on a definite integral, it is usually best to find the entire antiderivative before evaluating with the limits.

$$\int_1^2 xe^x \, dx \qquad\qquad \text{Let } u = x \text{ and } dv = e^x \, dx, \text{ so}$$
$$\qquad\qquad\qquad\qquad\qquad du = dx \text{ and } v = e^x.$$

$$\int_1^2 xe^x \, dx = xe^x \Big|_1^2 - \int_1^2 e^x \, dx$$
$$= (xe^x - e^x)\Big|_1^2$$
$$= 2e^2 - e^2 - (1e^1 - e^1)$$
$$= e^2.$$

1. $\int \dfrac{\ln x}{x^2}\,dx =$

    (A) $\dfrac{\ln x}{x^2} + C$

    (B) $\dfrac{\ln x}{x^2} - \dfrac{1}{x} + C$

    (C) $-\dfrac{\ln x}{x} - \dfrac{1}{x} + C$

    (D) $\dfrac{\ln x}{x} - \dfrac{1}{x} + C$

    (E) $-\dfrac{\ln x}{x} + \dfrac{1}{x^2} + C$

2. $\int \sin^{-1} x\,dx =$

    (A) $-x\sin^{-1}x - \sqrt{1 - x^2} + C$

    (B) $x\sin^{-1}x + \sqrt{x^2 - 1} + C$

    (C) $x\sin^{-1}x - \sqrt{x^2 - 1} + C$

    (D) $x\sin^{-1}x + \sqrt{1 - x^2} + C$

    (E) None of the above

3. Find the area enclosed by the graphs of $y = x\sin x$, $y = x - \pi$, and the y-axis.

    (A) $\dfrac{\pi^2}{2} + \pi$

    (B) $\dfrac{\pi^2}{2} - \pi$

    (C) $\pi - \dfrac{3}{2}\pi^2$

    (D) $\pi + \dfrac{3}{2}\pi^2$

    (E) π^2

Need More Help With . . .

Antidifferentiation by parts?

See . . .

Calculus, Section 6.3

Antidifferentiation by Simple Partial Fractions

AP* Objective:

Techniques of antidifferentiation

• Antiderivatives by substitution of variables (including change of limits for definite integrals), parts, and simple partial fractions (nonrepeating linear factors only)

Big Picture

Antidifferiation by partial fractions is strictly an AP* Calculus BC topic. It is another advanced technique of integration. Although it does not often appear in the Free Response portion of the AP* exam, it is a regularly tested item in the Multiple Choice section.

Content and Practice

As you learn the various techniques of integration, you should become skilled at deciding which technique must be applied in each situation. This process takes much practice. One clue for when the partial fraction method may be required is when an integrand's denominator can be factored into two or more different linear factors.

Since the AP* Calculus course limits the cases to nonrepeated linear factors, the general process is to break a single fraction into the sum of numerous fractions, each with a constant numerator and a linear denominator. Each individual fraction is then integrated, resulting in a sum or difference of natural logarithm antiderivatives. You need to know the rules for simplifying logarithms, since you may have to reconcile your answer to multiple choice options that may look slightly different from your result.

Study the example below, then try the following problems.

$$\int \frac{dw}{w^2 - 2w - 8}$$

$$\frac{1}{w^2 - 2w - 8} = \frac{1}{(w - 4)(w + 2)}$$

$$\frac{1}{(w - 4)(w + 2)} = \frac{A}{w - 4} + \frac{B}{w + 2}$$

Now multiply both sides of the equation by $(w - 4)(w + 2)$ to obtain $1 = A(w + 2) + B(w - 4)$. Some books present solving for A and B by setting up a system of equations, but a simple substitution method is somewhat more efficient.

Let $w = 4$, and the equation becomes

$$1 = A(4 + 2) + B(4 - 4)$$

so $A = \frac{1}{6}$ since B drops out.

Similarly, let $w = -2$, and the equation becomes

$$1 = A(-2 + 2) + B(-2 - 4)$$

so $B = -\frac{1}{6}$.

Thus

$$\frac{1}{(w - 4)(w + 2)} = \frac{\frac{1}{6}}{w - 4} + \frac{\frac{-1}{6}}{w + 2}$$

and the original integral can be rewritten as

$$\int \frac{dw}{w^2 - 2w - 8} = \frac{1}{6} \int \frac{dw}{w - 4} + \frac{-1}{6} \int \frac{dw}{w + 2}$$

$$= \frac{1}{6} \ln|w - 4| - \frac{1}{6} \ln|w + 2| + C$$

Why are there absolute value bars on the final answer? Are they always necessary?

Decompose the following fractions by the partial fraction method. You can check your answer by adding the resulting fractions or by using a computer algebra system and applying an *Expand* command to the original problem.

1. $\dfrac{8}{2x^2 - 3x - 2}$

2. $\dfrac{7x - 1}{x^2 + 4x - 21}$

1. $\int \dfrac{4}{x^2 + 8x + 15}\, dx =$

    (A) $2(\ln|x + 3| + \ln|x + 5|) + C$

    (B) $2(\ln|x + 3| - \ln|x + 5|) + C$

    (C) $2(\ln|x + 5| - \ln|x + 3|) + C$

    (D) $4\ln(x^2 + 8x + 15) + C$

    (E) $4(-x^{-1} + 4x^2 + 15x) + C$

2. Which is the solution to the initial value problem $\dfrac{dy}{dx} = \dfrac{3x + 1}{x^2 + x}$ and $y(1) = \ln 8$.

    (A) $y = \ln\left|4x + (x + 1)^2\right|$

    (B) $y = \ln|x| - 2 \cdot \ln|x + 1| + \ln 32$

    (C) $y = \ln\left|\dfrac{(x + 1)^3}{x}\right|$

    (D) $y = \dfrac{3}{2}\ln\left|x^2 + x\right| + \dfrac{1}{2}\ln 8$

    (E) $y = \ln\left|2x(x + 1)^2\right|$

Need More Help With . . .

Properties of logarithms?

Decomposition of fractions?

See . . .

Precalculus, Sections 3.3 and 3.4

Precalculus, Section 7.4

Calculus, Section 6.5 and 8.4

Improper Integrals

AP* Objective: Techniques of antidifferentiation
                            • Improper integrals (as limits of definite integrals)

Big Picture

Drawing together many of your calculus skills, *improper integrals* require recognition of discontinuities, skillful use of integration techniques, and limits sometimes involving L'Hôpital's Rule. They also lay a foundation for some of your work with converging and diverging series. It is an AP* Calculus BC only topic that often shows up on the Multiple Choice portion of the exam.

Content and Practice

A definite integral is improper if it has either an infinite limit of integration or any value from the lower limit of integration to the upper limit that causes an infinite discontinuity in the integrand. Proper form and use of limits is crucial to success with improper integrals. Any limit of integration value that causes an integral to be improper must be replaced with a variable and a limit approaching that value. You may also need to use L'Hôpital's Rule to evaluate some of the limits.

Study the following examples, then try the following problems.

- $\int_1^5 \dfrac{1}{\sqrt{x-1}}\, dx$ is improper because the lower limit 1 causes division by 0 in the integrand. So we rewrite the integral as

$$\lim_{a \to 1^+}\int_a^5 \frac{1}{\sqrt{x-1}}\, dx.$$

- Notice that a approaches 1 from above since the interval of integration is $[1, 5]$.

- $\int_0^7 \dfrac{1}{x-2}\, dx$ is improper since at $x = 2$, the integrand has an asymptote. This problem must be split into two integrals with limits:

$$\lim_{a \to 2^-}\int_0^a \frac{1}{x-2}\, dx + \lim_{a \to 2^+}\int_a^7 \frac{1}{x-2}\, dx.$$

Rewrite each integral in proper form using limits. *Do not integrate.*

1. $\int_0^\infty \tan^{-1} x \, dx$

2. $\int_0^5 \dfrac{1}{e^x - 1} \, dx$

3. $\int_0^6 \dfrac{dx}{\sqrt[3]{4 - x^2}}$

Improper integrals are also used to develop a small "library" of known convergent and divergent integrals. Using direct or limit comparison tests, these integrals can then be used to determine the convergence or divergence of more difficult integrals. Although too long to redevelop in this context, you should know that

- $\int_0^1 \dfrac{1}{x^p} \, dx$ converges for $0 < p < 1$ and diverges for $p \geq 1$.

- $\int_1^\infty \dfrac{1}{x^p} \, dx$ diverges for $0 < p \leq 1$ and converges for $p > 1$.

You also need to know the direct comparison test and may find the limit comparison test helpful.

Direct Comparison Test

Let f and g be continuous on $[a, \infty)$ with $0 \leq f(x) \leq g(x)$

for all $x \geq a$. Then

1. $\int_a^\infty f(x) \, dx$ converges if $\int_a^\infty g(x) \, dx$ converges.

2. $\int_a^\infty g(x) \, dx$ diverges if $\int_a^\infty f(x) \, dx$ diverges.

Limit Comparison Test

If the positive functions f and g are continuous on $[a, \infty)$ and if

$$\lim_{x \to \infty} \frac{f(x)}{g(x)} = L, \, 0 < L < \infty,$$

then

$\int_a^\infty f(x) \, dx$ and $\int_a^\infty g(x) \, dx$ both converge or both diverge.

Using the limit or direct comparison tests determine if the following integrals converge or diverge.

4. $\displaystyle\int_1^\infty \frac{dx}{1+x^3}$ 5. $\displaystyle\int_1^\infty \frac{dx}{x^2-0.9}$

6. $\displaystyle\int_0^1 \frac{dx}{1+\sqrt{x}}$ 7. $\displaystyle\int_1^\infty \frac{dx}{\ln x}$

Additional Practice

1. Find the first quadrant area under $y = \dfrac{3}{2e^{x/2}}$.

   (A) ∞ (B) $\dfrac{3}{2}e$ (C) 3

   (D) $\dfrac{3}{2}$ (E) $\dfrac{3}{4}$

2. $\displaystyle\int_2^\infty \frac{dx}{x^2+5x+6}$

   (A) 0 (B) 1 (C) $\ln 20$

   (D) $\ln\dfrac{5}{4}$ (E) ∞

3. $\displaystyle\int_0^1 x \ln x \, dx =$

   (A) -1 (B) $-\dfrac{1}{2}$ (C) $-\dfrac{1}{4}$

   (D) $-\infty$ (E) ∞

Need More Help With . . .   *See* . . .

L'Hôpital's Rule?   *Calculus*, Section 8.2

Improper integrals?   *Calculus*, Section 8.4

Initial Value Problems

Applications of antidifferentiation
• Finding specific antiderivatives using initial conditions, including applications to motion along a line

Big Picture

An equation containing a derivative dy/dx is called a *differential equation*. The general solution to the differential equation is the family of functions $y = f(x)$ that satisfy the differential equation. If in addition to the derivative dy/dx we are also given a point (x_0, y_0) that satisfies the function $y = f(x)$, we have an *initial value problem*. The value $y_0 = f(x_0)$ is called an initial value. The solution to the initial value problem is a function $y = f(x)$ that satisfies the differential equation and also contains the initial value. This solution is called a *particular solution*.

Content and Practice

The general solution to a differential equation is often found by integrating the derivative. This integral will include a constant of integration. The constant of integration can be found by substituting the initial value into the general solution. Once the constant of integration is found, we have the particular solution.

1. Find the solution to the initial value problem
    $dy/dx = \cos 2x$, $y(0) = 1$.

    Solution: The general solution is found by antidifferentiation.

    $$y = \int \cos 2x \, dx = \frac{1}{2} \sin 2x + C$$

    The constant of integration is found by substituting the initial value:

    $$1 = \frac{1}{2} \sin (2 \cdot 0) + C, \text{ so } C = 1$$

    The particular solution is

    $$y = \frac{1}{2} \sin 2x + 1.$$

2.   An arrow is fired straight up into the air with an initial velocity of $v(0) = 155$ feet/second.

(A) If the acceleration is $dv/dt = -32$ feet/second^2, find the velocity as a function of time.

(B) If the arrow is fired from an initial height of 5 feet, find the height as a function of time.

Additional Practice

1.   At time $t \geq 0$, the acceleration of a particle moving on the x-axis is $a(t) = t + \cos t$. At $t = 0$, the velocity of the particle is -3. For what value of t will the velocity of the particle be zero?
(A) 2.057   (B) 2.713   (C) 2.954
(D) 3.720   (E) 3.886

2.   The acceleration of a particle moving along the x-axis is $a = \cos(t)$. If $v(0) = 0$ and $x(0) = 1$, find the position $x(t)$ of the particle.

3.   If f is the antiderivative of $g(x) = \dfrac{x^3}{1 + x^5}$ such that $f(1) = 0$, then $f(5) = \underline{\quad}$.

At first this may appear to be a simple initial value problem in which you solve for $y = f(x)$ and then find $f(5)$. However, solving for $y = f(x)$ is difficult. In fact, it was not intended that students solve for $y = f(x)$.

Instead, define a function $f(x) = \int_1^x \frac{t^3}{1 + t^5} dt$. By the Fundamental Theorem, $f'(x) = \frac{x^3}{1 + x^5}$, so f is an antiderivative of $\frac{x^3}{1 + x^5}$ and $f(1) = 0$. Then $f(5) = \int_1^5 \frac{t^3}{1 + t^5} dt$, which can be evaluated with a calculator. Find $f(5)$.

Need More Help With . . .	See . . .
Initial value problems?	*Calculus,* Section 6.1
Particle motion problems?	*Calculus,* Sections 3.4 and 7.1

Separable Differential Equations

AP* Objective:

Applications of antidifferentiation
- Solving separable differential equations and using them in modeling; in particular, studying the equation $y' = ky$ and exponential growth
- Solving logistic differential equations and using them in modeling

Big Picture

A differential equation of the form $dy/dx = g(x)h(y)$ is said to be *separable*. We solve the differential equation by dividing both sides of the equation by $h(y)$ and then integrating both sides of the resulting equation.

$$\int \frac{1}{h(y)}\, dy = \int g(x)\, dx.$$

Content and Practice

The differential equation $\dfrac{dy}{dt} = ky$ is a separable differential equation that describes exponential growth (k positive) or decay (k negative). The general solution follows.

$$\int \frac{1}{y}\, dy = \int k\, dt \qquad \text{Separate variables.}$$

$$\ln|y| = kt + C \qquad \text{Integrate both sides.}$$

$$|y| = e^{kt+C} \qquad \text{Solve for } y.$$

$$|y| = e^C e^{kt} \qquad \text{Properties of exponents}$$

$$|y| = y_0 e^{kt} \qquad e^c = y_0$$

The value of y_0 can be found from initial conditions. We may need to find the value of k from another given point on the curve $y = f(t)$.

1. A bacteria colony with population y grows according to the differential equation $dy/dt = 2.3y$. There are 2000 bacteria initially.

   (A) Find the equation for population y as a function of time.

   (B) Find the number of bacteria at time $= 7$.

The differential equation $dP/dt = kP(M - P)$ is called a *logistic differential equation*. The equation describes the growth rate of a population P with carrying capacity M. Calculus AB students are not expected to solve this differential equation as it requires partial fractions, a skill unique to the BC curriculum. Recognizing that the equation represents logistic growth, and finding carrying capacity and k by algebraic methods is expected of all students. The solution is

$$\int \frac{1}{P(M - P)}\, dp = \int k\, dt \qquad \text{Separate variables}$$

$$\frac{1}{M}\int \left(\frac{1}{P} + \frac{1}{M - P}\right) dt = \int k\, dt \qquad \text{Partial fractions}$$

$$\ln |P| - \ln |M - P| = k\, Mt + C \qquad \text{Integrate both sides.}$$

$$\ln \left|\frac{M - P}{P}\right| = -kMt - C \qquad \text{Properties of logarithms}$$

$$\frac{M - P}{P} = e^{-kMt - C} \qquad \text{Definition of natural log with } O < P < M$$

$$\frac{M}{P} - 1 = (e^{-C})e^{-kt} = Ae^{-kMt} \qquad \text{Properties of exponents}$$

$$P = \frac{M}{1 + Ae^{-kMt}}$$

Notice $\lim_{t \to \infty} P(t) = M$ since the Ae^{-kMt} term goes to 0 as t goes to infinity. Thus, if by observation, a student can identify M and k in the differential equation, most of the critical information has been determined. For example:

$\frac{dP}{dt} = 3P - 0.003P^2$ can be factored into $\frac{dP}{dt} = 0.003P(1000 - P)$. By comparison to the standard differential equation we can determine $M = 1000$ and $k = 0.003$.

2. Consider the logistic differential equation $\dfrac{dP}{dt} = \dfrac{P}{3}\left(1 - \dfrac{P}{12}\right)$.

(A) Write the differential equation in the standard form shown above.

(B) If $P(0) = 5$, what is $\lim\limits_{t\to\infty} P(t)$?

Additional Practice

1. Find $y(t)$ if $\dfrac{dy}{dt} = \dfrac{y}{3}\left(1 - \dfrac{t}{4}\right)$ and $y(0) = 2$.

2. Population y grows according to the equation $dy/dt = ky$, where k is a constant and t is measured in years. If the population doubles every 8 years, which of the following could be the value of k?
   (A) 0.087   (B) 0.349   (C) 0.799
   (D) 1.071   (E) 1.152

3. Solve the differential equation $dy/dx = -2xy$ and $y(1) = 4$.

4. Without integrating, find the carrying capacity for a population growth rate modeled by $\dfrac{dp}{dt} = 6P - 0.012\,P^2$.

   (A) 500   (B) 50   (C) 0.012
   (D) 0.002   (E) None of these

Need More Help With . . .   *See . . .*

Separable differential equations?   *Calculus,* Section 6.4

Exponential growth and decay?   *Calculus,* Section 6.4

Logistic growth?   *Calculus,* Section 6.5

Partial fractions?   *Calculus,* Section 6.5

Numerical Approximations to Definite Integrals

Numerical approximations to definite integrals

Big Picture

In theory, *definite integrals* can be used in a wide variety of real-world applications including area, volume, work, distance, and velocity. In practice, these definite integrals may be difficult or impossible to evaluate directly. This is why it is helpful to know methods that closely approximate definite integrals. Some of these methods include left-hand, right-hand, and midpoint Riemann sums as well as the Trapezoidal Rule.

Content and Practice

The section in this workbook titled "Riemann Sums" described left-hand, right-hand, and midpoint rectangular methods to approximate definite integrals. You may want to review that section before continuing. Each of these three methods uses rectangles to approximate a definite integral. Another geometric shape that may give a more accurate approximation is the trapezoid.

Trapezoidal Rule

If the interval $[a, b]$ is partitioned into n equal subintervals of width $h = \dfrac{b - a}{n}$, the Trapezoidal Rule to approximate the definite integral of $y = f(x)$ on the interval $[a, b]$ is

$$\int_a^b f(x)\, dx \approx \frac{h}{2} [y_0 + 2y_1 + 2y_2 + \cdots + 2y_{n-1} + y_n].$$

Note that the trapezoidal approximation is equal to the average of the left- and right-hand rectangular approximation methods.

If the subintervals are not of equal length, you must find the area of each trapezoid separately and sum the areas to approximate the definite integral. The area of a single trapezoid is

$$\frac{h}{2}[y_i + y_{i+1}].$$

1.   Consider the area bounded by $f(x) = 4 - x^2$ and the x-axis.

   (A) Approximate this area using the Trapezoidal Rule with $n = 4$.

   (B) Find the left- and right-hand Riemann sums using $n = 4$ and verify that the average of these two approximations gives the same result as the Trapezoidal Rule.

2.   The table below shows the velocity of a remote-controlled toy car as it traveled down a hallway for 10 seconds.

Time (sec)	0	1	2	3	4	5	6	7	8	9	10
Velocity (in./sec)	0	6	10	16	14	12	18	22	12	4	2

Using the Trapezoidal Rule, estimate the distance traveled by the car. Use 10 subintervals of length 1.

Additional Practice

1.   The graph of $y = f(x)$ is shown. Use the Trapezoidal Rule with $n = 4$ to estimate the area bounded by the graph of $y = f(x)$ and the x-axis on the interval $[-2, 2]$.

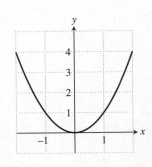

2. The function f is continuous on the closed interval $[1, 7]$ and has the values given in the table

x	1	4	6	7
$f(x)$	10	20	40	30

Using the subintervals $[1, 4]$, $[4, 6]$, and $[6, 7]$, what is the trapezoidal approximation of $\int_a^b f(x)\, dx$?

(A) 110   (B) 120   (C) 130   (D) 140   (E) 150

Need More Help With . . .   ***See . . .***

Rectangular approximation   *Calculus,* Section 5.1
methods?

Trapezoidal Rule?   *Precalculus,* Section 5.5

Concept of Series

Concept of series

Big Picture

A *series* is defined as a sequence of partial sums, and *convergence* is defined in terms of the limit of the sequence of partial sums. Technology can be used to explore convergence or divergence.

Content and Practice

An infinite series is an expression of the form

$$a_1 + a_2 + a_3 + \cdots + a_n + \cdots = \sum_{k=1}^{\infty} a_k.$$

The partial sums of the series form a sequence

$$s_1 = a_1$$
$$s_2 = a_1 + a_2$$
$$s_3 = a_1 + a_2 + a_3$$
$$\vdots$$
$$s_n = \sum_{k=1}^{n} a_k$$

of real numbers. If the sequence of partial sums converges to a value S, then the infinite series converges to the sum S, which can be written $\sum_{k=1}^{\infty} a_k = S$.

1. The infinite geometric series $\sum_{k=1}^{\infty}\left(\frac{1}{2}\right)^k$ converges to 1. A table and graph of the sequence of partial sums provides evidence for convergence.

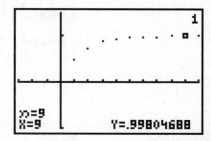

2. Make a table and graph of the sequence of partial sums for each of the following series. Use the sequence of partial sums to predict whether the infinite series converges or diverges. If the infinite series converges, estimate the value to which it converges.

(A) $\displaystyle\sum_{k=1}^{\infty} \frac{1}{k^2}$

(B) $\displaystyle\sum_{k=1}^{\infty} \frac{k}{k+2}$

Additional Practice

1. Use a graph of the sequence of partial sums to predict whether the infinite series

$$\sum_{k=1}^{\infty} \frac{(-1)^{k+1}}{k}$$

converges or diverges.

2. Use a table of the sequence of partial sums to estimate the value to which the infinite series $\sum_{k=0}^{\infty}\left(\frac{-1}{3}\right)^k$ converges.

(A) 0.33 (B) 0.50 (C) 0.75 (D) 1.0 (E) 1.33

Need More Help With . . .   See . . .

Infinite series?   *Calculus,* Section 9.1

Geometric, Harmonic, and Alternating Series

AP* Objective:

Series of constants
- Motivating examples, including decimal expansion
- Geometric series with applications
- The harmonic series
- Alternating series with error bound

Big Picture

Geometric, harmonic, and alternating series have very specific forms and properties. If you can identify which type of series you are dealing with, it is easy to decide questions about convergence.

Content and Practice

A geometric series takes the form

$$a + ar + ar^2 + ar^3 + \cdots + ar^{n-1} + \cdots = \sum_{n=1}^{\infty} ar^{n-1}.$$

If $|r| < 1$, the series converges to $\dfrac{a}{1-r}$; otherwise, the series diverges.

The harmonic series takes the form

$$1 + \frac{1}{2} + \frac{1}{3} + \frac{1}{4} + \cdots + \frac{1}{n} + \cdots = \sum_{n=1}^{\infty} \frac{1}{n}.$$

The harmonic series diverges.

An alternating series takes the form

$$u_1 - u_2 + u_3 - u_4 + \cdots = \sum_{n=1}^{\infty} (-1)^{n+1} u_n,$$

where each u_n is positive. If $u_n \geq u_{n+1}$ for all $n \geq N$ for some integer N and if $\lim_{n \to \infty} u_n = 0$, the series converges. If the series converges, the truncation error for the nth partial sum is u_{n+1} and has the same sign as the first unused term.

1. Which of the following series converge?

   I. $\displaystyle\sum_{n=1}^{\infty}\left(\frac{4}{3}\right)^{n}$ II. $\displaystyle\sum_{n=1}^{\infty}\frac{\cos n\pi}{n}$ III. $\displaystyle\sum_{n=1}^{\infty}\frac{1}{n}$

   (A) None   (B) II only   (C) III only
   (D) I and II only   (E) I and III only

2. Show that $1 - \dfrac{1}{3!}$ approximates $\displaystyle\sum_{n=0}^{\infty}\frac{(-1)^{n}}{(2n+1)!}$ with an error of less than $\dfrac{1}{100}$. Is the approximation an overestimate or an underestimate?

3. $\displaystyle\sum_{n=1}^{\infty}5\left(\frac{2}{3}\right)^{n-1} = $ _____.

Additional Practice

1. What is the sum of the infinite geometric series
   $$\frac{3}{2} + \frac{9}{16} + \frac{27}{128} + \frac{81}{1024} + \cdots?$$

   (A) $\dfrac{7}{4}$ (B) 2   (C) $\dfrac{12}{5}$

   (D) $\dfrac{5}{2}$ (E) There is no finite sum.

2. Consider the infinite series $\displaystyle\sum_{n=1}^{\infty}4\left(\frac{-1}{3}\right)^{n-1}$.

   (A) What is the sum of this series?

   (B) How many terms are required in the partial sum to approximate the sum of the infinite series to within 0.001?

Need More Help With . . .   See . . .

Geometric series?	*Calculus*, Section 9.1
Harmonic and alternating series?	*Calculus*, Section 9.5

Integral Test, Ratio Test, and Comparison Test

Series of constants

- Terms of series as areas of rectangles and their relationship to improper integrals, including the Integral Test and its use in testing the convergence of p-series
- The Ratio Test for convergence and divergence
- Comparing series to test for convergence and divergence

Big Picture

Integral Test: Suppose that $a_n = f(n)$, where f is a continuous, positive, decreasing function of x for all $x \geq N$ (N a positive integer). Then the series $\sum_{n=N}^{\infty} a_n$ and the integral $\int_N^{\infty} f(x)\, dx$ either both converge or both diverge.

Ratio Test: Let $\sum a_n$ be a series with all positive terms and with

$$\lim_{n \to \infty} \frac{a_{n+1}}{a_n} = L.$$

- If $L < 1$, the series converges.
- If $L > 1$, the series diverges.
- If $L = 1$, the test is inconclusive.

Comparison Test: Let $\sum a_n$ be a series with no negative terms.

- $\sum a_n$ converges if there is a convergent series $\sum c_n$ with $a_n \leq c_n$ for all $n > N$ for some integer N.
- $\sum a_n$ diverges if there is a divergent series $\sum d_n$ of nonnegative terms with $a_n \geq d_n$ for all $n > N$ for some integer N.

Content and Practice

1. Does $\sum_{n=1}^{\infty} \frac{1}{n^2}$ converge or diverge?

    Solution: Consider the corresponding improper integral:

    $$\int_1^{\infty} \frac{1}{x^2}\, dx = \lim_{b \to \infty} \int_1^b \frac{1}{x^2}\, dx$$

    $$= \lim_{b \to \infty} \left(1 - \frac{1}{b}\right) = 1$$

    Since the improper integral converges, it follows that $\sum_{n=1}^{\infty} \frac{1}{n^2}$ converges.

    It can be shown by the Integral Test that any series of the form $\sum_{n=1}^{\infty} \frac{1}{n^p}$ converges if $p > 1$ and diverges if $p \leq 1$. A series of the form $\sum_{n=1}^{\infty} \frac{1}{n^p}$ is called a *p-series*.

2. Does the series $\sum_{n=1}^{\infty} \frac{3^n}{n!}$ converge or diverge?

3. Does the series $\sum_{n=1}^{\infty} \frac{1}{n^2 + n + 1}$ converge or diverge?

Additional Practice

1. Which of the following series converge?

    I. $\displaystyle\sum_{n=1}^{\infty} \frac{1}{\sqrt{n}}$ II. $\displaystyle\sum_{n=1}^{\infty} \frac{n!}{n^n}$ III. $\displaystyle\sum_{n=1}^{\infty} \frac{1}{n^3 + 1}$

    (A) I only   (B) I and II only   (C) I and III only

    (D) II and III only   (E) III only

Need More Help With . . .   **See . . .**

Ratio Test or Comparison Test?   *Calculus*, Section 9.4

Integral Test?   *Calculus*, Section 9.5

Taylor Polynomials

Taylor series
• Taylor polynomial approximation with graphical demonstration of convergence

Big Picture

Early in the course we approximated a function near a given point by using a tangent line. The problem was that it quickly became very inaccurate for most functions. We find that as we increase the number of terms and degree of the approximating polynomial, the polynomial can approximate a nonpolynomial function with greater accuracy. In the study of Taylor series we examine these concepts, look at construction of these polynomials, explore domains over which they approximate other functions, and consider the error involved in using polynomials for approximation. This is a topic required only in the AP* Calculus BC curriculum.

Content and Practice

Graphed below are the functions $y = \cos x$ and an increasing number of the terms of the polynomial $P(x) = 1 - \dfrac{x^2}{2!} + \dfrac{x^4}{4!} - \dfrac{x^6}{6!} + \cdots$. Notice how the polynomial begins to approximate the cosine function on a wider and wider domain.

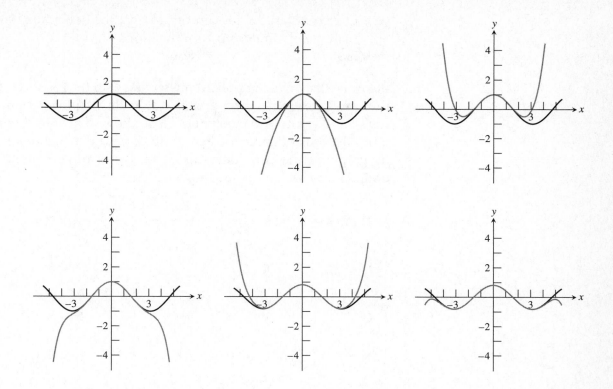

We can show that the infinite polynomial series converges to the cosine function for all real numbers. If we truncate the series, the only place it will exactly equal the cosine is at the original point of tangency, the center of the series. But a higher-order polynomial will certainly provide greater accuracy over a wider domain than a linear approximation.

1. Examine the relationship between $y = \ln(x - 1)$ and an increasing number of terms of the polynomial

$$P(x) = (x - 2) - \frac{(x - 2)^2}{2} + \frac{(x - 2)^3}{3} - \frac{(x - 2)^4}{4} + \cdots.$$

Graph $y = \ln(x - 1)$ and $P(x)$ together on the same screen. First use only the linear factor of $(x - 2)$. Then add the quadratic term and examine the graph again. Next add the cubic term, and so on.

(A) Where is the polynomial exactly equal to the natural logarithm function?

(B) Does adding more terms appear to create a polynomial that "fits" the curve better?

(C) Over how large a domain can you get the polynomial to approximate the natural logarithm function?

2. Below is a list of Taylor polynomials along with the functions that match the series. Examine their graphs gradually by plotting an increasing number of terms of the polynomial. See how soon you recognize which nonpolynomial function is being approximated. Check your conjecture by graphing the two functions at the same time. Comment on the domain over which the functions match.

(A) $P(x) = x + \dfrac{(x-1)^2}{2!} + \dfrac{(x-1)^3}{3!} + \dfrac{(x-1)^4}{4!} + \cdots$

(B) $P(x) = x - \dfrac{x^3}{3} + \dfrac{x^5}{5} - \dfrac{x^7}{7} + \cdots$

(C) $P(x) = x - \dfrac{x^3}{3!} + \dfrac{x^5}{5!} - \dfrac{x^7}{7!} + \cdots$

(D) $\sum_{n=1}^{\infty} \dfrac{x^{n-1}}{(-2)^{n-1}}$

(E) $P(x) = x + \dfrac{x^3}{3} + \dfrac{2x^5}{15} + \dfrac{17x^7}{315} + \cdots$

(A) $y = \dfrac{2}{2+x}$ (B) $y = \sin x$ (C) $y = \tan x$

(D) $y = \tan^{-1} x$ (E) $y = e^{x-1}$

Additional Practice

⊞ 1. The Taylor polynomial

$$P(x) = \dfrac{x}{3} + \dfrac{x^2}{9} + \dfrac{x^3}{27} + \dfrac{x^4}{81} + \cdots$$

is used to approximate the function $f(x) = \dfrac{x}{3-x}$. How many terms of the polynomial are needed so that the error between the polynomial approximation and the function at $x = 0.4$ is less than $\dfrac{1}{1000}$?

(A) 1   (B) 2   (C) 3   (D) 4   (E) 5

2. The Taylor polynomial

$$P(x) = 1 - 2x^2 + \frac{2x^4}{3} - \frac{4x^6}{45} + \frac{2x^8}{315} - \frac{4x^{10}}{14175}$$

approximates the function $f(x) =$

(A) $\cos 2x$.　　　　(B) $\sin 2x$.　　　　(C) $\cos x^2$.

(D) $1 - \sin x^2$.　　(E) $\cos x$.

Need More Help With . . .

Taylor polynomials?

See . . .

Calculus, Section 9.2

Maclaurin and Taylor Series

Taylor series
- Maclaurin series and general Taylor series centered at $x = a$
- Maclaurin series for the functions e^x, $\sin x$, $\cos x$, and $\dfrac{1}{1-x}$

Big Picture

Maclaurin and Taylor series are the entire focus of Chapter 9 in the *Calculus* text. Familiarity with the four series listed above is necessary, as they represent the most common series used throughout the course. You can almost always count on a Taylor series question on the Free Response portion of the AP* exam, and a series question or two among the Multiple Choice problems. Series are also important building blocks for further study beyond the scope of this course.

Content and Practice

Once a visual understanding of a series approximating a function is established, we move into actually learning to produce those series. We are extending the idea of a linear approximation at a point of tangency to a higher-degree polynomial. The original point of tangency is the center of the series. Students often miss the simple idea that a Maclaurin series is just a special case of a Taylor series. A Maclaurin series is centered at 0, while a Taylor series can be centered at any value $x = a$ in the domain of the original function.

The general form of a Taylor series for a function f centered at $x = a$ is:

$$\sum_{k=0}^{\infty} \frac{f^{(k)}(a)}{k!}(x-a)^k = f(a) + f'(a)(x-a) + \frac{f''(a)}{2!}(x-a)^2 + \cdots + \frac{f^k(a)}{k!}(x-a)^k + \cdots$$

The infinite series requires the function to have derivatives of all orders. If we acknowledge that our approximation will have less accuracy than an infinite series, a Taylor polynomial of finite order n can be used. It is simply a partial sum from the series above. The order of a series is the highest order of the derivative of f used in forming the series.

$$\sum_{k=0}^{n} \frac{f^{(k)}(a)}{k!}(x-a)^k = f(a) + f'(a)(x-a) + \frac{f''(a)}{2!}(x-a)^2 + \cdots + \frac{f^n(a)}{n!}(x-a)^n$$

Notice that the first two terms of the series form the linearization we worked with through much of the course.

The development of why the formula works is presented more thoroughly in the *Calculus* text. The Maclaurin series for $f(x) = e^x$ is one of the easiest to develop from scratch. We simply need to evaluate its derivatives at $a = 0$. Let's review the process.

1. Evaluate the function and its successive derivatives at the center $a = 0$.

    $f(0) = e^0 = 1$ and since the derivative of e^x
    is still e^x, the value of all derivatives at 0 is 1.

2. Substitute the value of the function, its derivatives, and a into the Taylor formula.

$$e^x = \sum_{k=0}^{\infty} \frac{f^{(k)}(0)}{k!}(x - 0)^k = f(0) + f'(0)(x - 0) + \frac{f''(0)}{2!}(x - 0)^2 + \cdots$$

$$+ \frac{f^k(0)}{k!}(x - 0)^k + \cdots$$

3. Remembering the conclusion in step 1, simplify the formula.

$$e^x = \sum_{k=0}^{\infty} \frac{(x)^k}{k!} = 1 + x + \frac{x^2}{2!} + \frac{x^3}{3!} + \cdots + \frac{x^k}{k!} + \cdots$$

This is the infinite Maclaurin series for e^x. It converges to e^x for all real numbers x.

To develop the Maclaurin series for $\frac{1}{1 - x}$, we could use the same procedure. But as you will recall from earlier in this workbook, this function has the form of a geometric series. It is significantly less work to recognize it has the form $S_\infty = \frac{a}{1 - r}$ where $a = 1$ and $r = x$. Then the series generates nicely to $\frac{1}{1 - x} = 1 + x + x^2 + x^3 + \cdots + x^n + \cdots$. Remember also that it converges only when $|r| < 1$, so the series only converges to the function on the interval $-1 < x < 1$.

In the previous section of this workbook, you were given the Maclaurin series for $y = \cos x$. Now try to develop it yourself by using the same process we used for e^x.

Additional Practice

1. Let g be a function that has derivatives of all orders. Assuming $g(0) = 4$, $g'(0) = 7$, $g''(0) = -2$, $g'''(0) = 5$, and $g^4(0) = -8$, write a fourth-degree Taylor polynomial for g centered at $x = 0$.

2. Determine the Maclaurin series for $f(x) = 1/x$ centered at $a = 1$.

3. For a given function $h(x)$, a 12th-order Taylor polynomial is written in ascending powers of x, so that the last term of the polynomial is $\frac{x^{12}}{3960}$. Which is the value of the 12th derivative of h at $x = 0$, $h^{12}(0)$?

   (A) 9!   (B) $3 \cdot 8!$ (C) 12!

   (D) 3960^2 (E) Cannot be determined

Need More Help With . . .

Taylor or Maclaurin series?

See . . .

Calculus, Section 9.2

Manipulating Taylor Series

AP* Objective:

Taylor series
- Formal manipulation of Taylor series and shortcuts to computing Taylor series, including substitution, differentiation, antidifferentiation, and the formation of new series from known series

Big Picture

Just as there are rules of acceptable processes in algebra (such as always keeping a balanced equation), there are rules to govern what can and cannot be done with Taylor series. Your goal is to understand those rules to work most efficiently and confidently with series.

Content and Practice

If we wanted to create a Taylor series for e^{2x} centered at $a = 0$, we would evaluate its derivatives at 0 and substitute into the Taylor series formula below. Showing just the first few steps reveals an interesting pattern.

$$\sum_{k=0}^{\infty} \frac{f^{(k)}(a)}{k!}(x-a)^k = f(a) + f'(a)(x-a) + \frac{f''(a)}{2!}(x-a)^2 + \cdots + \frac{f^k(a)}{k!}(x-a)^k + \cdots$$

$$\begin{aligned}
f(x) &= e^{2x} &\text{so} \quad f(0) &= e^0 = 1 \\
f'(x) &= 2e^{2x} &\text{so} \quad f'(0) &= 2e^0 = 2 \\
f''(x) &= 4e^{2x} &\text{so} \quad f''(0) &= 4e^0 = 4 \\
f'''(x) &= 8e^{2x} &\text{so} \quad f'''(0) &= 8e^0 = 8
\end{aligned}$$

So a third-order Taylor polynomial for e^{2x} centered at $a = 0$ is

$$e^{2x} \approx 1 + 2x + \frac{4}{2!}x^2 + \frac{8}{3!}x^3,$$

or, rewriting,

$$e^{2x} \approx 1 + 2x + \frac{(2x)^2}{2!} + \frac{(2x)^3}{3!}.$$

Compare this to the series for e^x,

$$e^x = \sum_{k=0}^{\infty} \frac{(x)^k}{k!} = 1 + x + \frac{x^2}{2!} + \frac{x^3}{3!} + \cdots + \frac{x^k}{k!} + \cdots.$$

Notice that x in the series for e^x has just been replaced by $2x$ to obtain a series for e^{2x}.

   The question that must now be answered is, "When is a replacement of variables into a known series allowed, and when is it not?" The simplest solution is to keep your attention focused on the center of each series. If the centers of the series are the same, a new series can be generated from a known series by substitution. Let's look at a couple of examples where substituting does and does not work.

Example 1: The series for $f(x) = \sin x$ centered at $a = 0$ is

$$\sin x = \sum_{n=0}^{\infty} (-1)^n \frac{x^{2n+1}}{(2n+1)!} = x - \frac{x^3}{3!} + \frac{x^5}{5!} - \cdots.$$

Example 2: The series for $g(x) = \sin 3x$ centered at $a = 0$ is

$$\sin 3x = \sum_{n=0}^{\infty} (-1)^n \frac{(3x)^{2n+1}}{(2n+1)!} = 3x - \frac{(3x)^3}{3!} + \frac{(3x)^5}{5!} - \cdots.$$

Example 3: The series for $h(x) = \sin(3x + 1)$ centered at $a = 0$ *is not*

$$\sin(3x+1) = \sum_{n=0}^{\infty} (-1)^n \frac{(3x+1)^{2n+1}}{(2n+1)!} = (3x+1) - \frac{(3x+1)^3}{3!} + \frac{(3x+1)^5}{5!} - \cdots.$$

This is a series for $\sin(3x + 1)$, but it is centered at $a = -1/3$. To understand why it is not centered at 0, think of $f'(0)$. $f'(x) = 3\cos(3x + 1)$ and $f'(0) = 3\cos 1$, but this value and successive derivatives of $\sin x$ at $x = 0$ could not possibly show up in the series by substituting only for x.

   We can also generate new series by differentiating or integrating. In the previous section, you were asked to develop from scratch the Maclaurin series for $\cos x$. We could also generate it by differentiating the series for $\sin x$.

$$\sin x = \sum_{n=0}^{\infty} (-1)^n \frac{x^{2n+1}}{(2n+1)!} = x - \frac{x^3}{3!} + \frac{x^5}{5!} - \cdots$$

$$\frac{d}{dx}\sin x = \frac{d}{dx}\sum_{n=0}^{\infty} (-1)^n \frac{x^{2n+1}}{(2n+1)!} = \frac{d}{dx}\left(x - \frac{x^3}{3!} + \frac{x^5}{5!} - \cdots \right)$$

$$\cos x = \sum_{n=0}^{\infty} (-1)^n \frac{x^{2n}}{(2n)!} = 1 - \frac{x^2}{2!} + \frac{x^4}{4!} - \cdots$$

A common mistake made by students when integrating a series is failing to account for the arbitrary constant. Consider the series for $f(x) = \dfrac{1}{2-x}$ centered at $a = 1$. If we write $f(x) = \dfrac{1}{1-(x-1)}$, we can write a geometric series with first term 1 and common ratio $(x-1)$:

$$\frac{1}{2-x} = 1 + (x-1) + (x-1)^2 + (x-1)^3 + \cdots.$$

Integrating both sides of the equation we get

$$-\ln(2-x) = x + \frac{(x-1)^2}{2} + \frac{(x-1)^3}{3} + \frac{(x-1)^4}{4} + \cdots + C$$

We now substitute the only point where the series exactly equals the function, at the center $x = 1$, to get $0 = 1 + 0 + 0 + 0 + \cdots + C$. So $C = -1$ and the series becomes

$$-\ln(2-x) = (x-1) + \frac{(x-1)^2}{2} + \frac{(x-1)^3}{3} + \frac{(x-1)^4}{4} + \cdots.$$

We have used one series and integration to develop a series for a new function centered in the same place. Had we not considered the constant, our new series would have been missing a term.

1.   The geometric series

$$\frac{1}{1+x^2} = \sum_{n=0}^{\infty}(-1)^n x^{2n} = 1 - x^2 + x^4 - x^6 + \cdots$$

can be integrated to produce a series for a new function. Find the function, the new series, a general term, and identify the center of the series.

2.   From the Maclaurin series for e^x, can substitution be used to generate a Maclaurin series for e^{2x-1}? Explain your answer in few sentences.

Additional Practice

1. Given the Maclaurin series

    $$e^{x^2} = 1 + x^2 + \frac{x^4}{2!} + \cdots + \frac{(x)^{2n}}{n!} + \cdots,$$

    find a series representation for xe^{x^2}. Write the first three terms and the general term.

2. Which is the center of the series $\displaystyle\sum_{n=0}^{\infty}(-1)^n \frac{(4x-1)^{2n}}{n!}$?

    (A) -4 (B) $-\frac{1}{4}$ (C) 0 (D) $\frac{1}{4}$ (E) 4

3. Let $P(x) = 7 - 3(x-4) + 5(x-4)^2 - 2(x-4)^3 + 6(x-4)^4$ be the fourth-degree Taylor polynomial for the function f about 4. Assume f has derivatives of all orders for all real numbers. What is the third-degree Taylor polynomial for $g(x) = \int_4^x f(t)\, dt$ about 4?

    (A) $7 - 3(x-4) + 5(x-4)^2 - 2(x-4)^3$

    (B) $7x - \dfrac{3(x-4)^2}{2} + \dfrac{5(x-4)^3}{3} - 28$

    (C) $-3 + 10(x-4) - 6(x-4)^2 + 24(x-4)^3$

    (D) $7x - \dfrac{3(x-4)^2}{2} + \dfrac{5(x-4)^3}{3}$

    (E) Cannot be determined

Need More Help With . . .

Manipulation of series?

See . . .

Calculus, Section 9.1

Power Series

AP* Objective:

Taylor series
• Functions defined by power series

Big Picture

Power series are the foundation upon which we build Maclaurin and Taylor series. They are also a good starting point for understanding the concept of series representing functions graphically and the idea of convergence. Power series flow directly from our work with geometric series.

Content and Practice

A power series is an expression of the form

$$\sum_{j=0}^{\infty} c_j x^j = c_0 + c_1 x + c_2 x^2 + \cdots + c_j x^j + \cdots.$$

All c_j's are real constants and may at times be 0. We say the series is centered at $x = 0$.

If we center the series somewhere other than zero, say $x = a$, we replace all x's with $(x - a)$:

$$\sum_{j=0}^{\infty} c_j (x - a)^j = c_0 + c_1(x - a) + c_2(x - a)^2 + \cdots + c_j(x - a)^j + \cdots.$$

Consider the power series

$$P(x) = 2 + 4x^2 + 8x^4 + 16x^6 + \cdots.$$

If we recognize it as an infinite geometric series, its sum will be $S_\infty = \dfrac{a}{1 - r}$. The first term is $a = 2$ and the common ratio is $r = 2x^2$, so the sum of the function is

$$f(x) = \frac{2}{1 - 2x^2}.$$

If a power series is geometric, we can determine an interval of convergence by setting $|r| < 1$. Unfortunately, not all power series are geometric, and in those cases it can be extremely difficult to determine an infinite sum and an interval of convergence.

Power series may also be differentiated or integrated term by term to produce new series to model different functions. Consider our previous example. On a certain domain,

$$\frac{2}{1 - 2x^2} = 2 + 4x^2 + 8x^4 + 16x^6 + \cdots.$$

Differentiating both sides of the equation produces

$$\frac{8x}{(1 - 2x^2)^2} = 8x + 32x^3 + 96x^5 + \cdots.$$

Notice the new series is not geometric. It would be very difficult to determine its sum. Similarly, if we knew the function and wanted to generate the series, because the function is not in the form of a sum of an infinite geometric series we would have an extremely difficult task before us.

1. Given the infinite geometric series $x - 3x^2 + 9x^3 - 27x^4 + \cdots$, identify a and r and determine the function represented by the sum of the series.

2. By writing it in the form $\frac{a}{1 - r}$, generate a power series for the function $f(x) = \frac{4}{2 - 3x}$ that is centered at each of the following values.

    (A) $x = 0$

    (B) $x = \frac{1}{3}$

Additional Practice

1. The seventh-order power series representation of $g(x) = \sin^{-1}x$ at $x = 0$ is $x + \dfrac{x^3}{6} + \dfrac{3x^5}{40} + \dfrac{5x^7}{112}$. Which is a power series representation of $f(x) = \dfrac{x}{\sqrt{1 - x^2}}$ at $x = 0$?

   (A) $1 + \dfrac{x^2}{2} + \dfrac{3x^4}{8} + \dfrac{5x^6}{16}$ (B) $\dfrac{x^2}{2} + \dfrac{x^4}{24} + \dfrac{x^6}{80} + \dfrac{5x^8}{896}$

   (C) $x + \dfrac{x^3}{2} + \dfrac{3x^5}{8} + \dfrac{5x^7}{16}$ (D) $x^{1/2} + x^{3/2} + x^{5/2} + x^{7/2}$

   (E) Cannot be determined

2. Determine the coefficient of the 12th-degree term of the power series representation of $h(x) = \dfrac{3}{1 + 2x^3}$ centered at $x = 0$.

   (A) $-3 \cdot 2^{12}$ (B) -48 (C) 0
   (D) 48 (E) $3 \cdot 2^{12}$

3. What is the interval of convergence of the power series $\displaystyle\sum_{n=0}^{\infty} \dfrac{(x - 3)^n}{2^{n+1}}$?

   (A) $1 < x < 5$ (B) $2 < x < 4$ (C) $-1 < x < 1$
   (D) $-2 < x < 2$ (E) $-\infty < x < \infty$

Need More Help With . . .

Power series?

See . . .

Calculus, Section 9.1

Radius and Interval of Convergence

Taylor series
• Radius and interval of convergence of a power series

Big Picture

Although a power series is useful for approximating functions, we must pay close attention to the domain over which the series converges to the function. Some power series converge for all real numbers, while some converge over a limited domain.

Content and Practice

While developing a graphical understanding of convergence, you should have visually determined that series converge to the functions they are modeling on intervals that can vary greatly. Some converge on $-1 < x < 1$, others converge on somewhat wider domains, and others converge for all real numbers. As we noted in the previous section, it is not terribly difficult to analytically find the interval of convergence for an infinite power series that is geometric. We identify the common ratio, r, and then solve the inequality $|r| < 1$. The conjunction obtained will be of the form $a < x < b$ and this is the interval over which the power series converges to the function it represents.

Outside this interval, we cannot rely on the series to provide an accurate approximation of the function. Just as the radius of a circle is the distance from its center to the circle, the radius of convergence is the distance from the center of the interval of convergence to one end of the interval. Another way to see the radius of convergence is to write the above conjunction in the form $|x - c| < R$, where c is the center of the series and R is the radius of convergence. If we differentiate or integrate a power series, the radius and center of convergence remain the same for the new series obtained.

1. Consider the infinite geometric series $\sum_{n=0}^{\infty}(x - 3)^n$.

   (A) Find the function represented by the series.

   (B) Find the interval of convergence.

   (C) Find the center of the series.

   (D) Find the radius of convergence.

2. If the interval of convergence of an infinite series is $-3 < x < 11$, which is its radius of convergence?

   (A) -3 (B) 4   (C) 7   (D) 11   (E) 14

3. If the interval of convergence of the series representing $f(x)$ is $\frac{3}{2} < x < \frac{11}{2}$, then the interval of convergence of the series representing $f'(x)$ is _____.

   If a power series is not geometric, then we can often use the Ratio Test to determine the interval of convergence. Recall the conditions of the Ratio Test:

Let $\sum a_n$ be a series such that $\lim\limits_{n \to \infty} \left| \dfrac{a_{n+1}}{a_n} \right| = L$.

If $L < 1$, the series converges.
If $L > 1$, the series diverges.
If $L = 1$, the test is inconclusive.

Let's examine the series defined by $\sum_{n=0}^{\infty} n(3x-5)^n$. If we expand it, the factor of n prevents the series from being geometric. The first few terms are $(3x-5) + 2(3x-5)^2 + 3(3x-5)^3 + \cdots$. We apply the Ratio Test to the general term.

$$\lim_{n \to \infty} \left| \frac{(n+1)(3x-5)^{n+1}}{n(3x-5)^n} \right| < 1$$

$$\lim_{n \to \infty} \left| \frac{n+1}{n}(3x-5) \right| < 1$$

$$\left| 1(3x-5) \right| < 1$$

$$-1 < 3x - 5 < 1$$

$$\frac{4}{3} < x < 2$$

Testing the endpoints in the series, we find that it fails to converge at both endpoints.

4. Apply the Ratio Test to find the interval of convergence of the series $\sum_{k=0}^{\infty} \dfrac{n(x-1)^n}{5^n}$.

Additional Practice

1. Which is the sum of the infinite power series $\sum_{n=1}^{\infty} \dfrac{(x-1)^n}{2^n}$?

    (A) $\dfrac{x-1}{3-x}$ (B) $\dfrac{2}{3-x}$ (C) $\dfrac{2}{1-x}$

    (D) $\dfrac{x-1}{2-x}$ (E) $\dfrac{x-1}{1+x}$

2. If $f(x)$ is an infinite power series with an interval of convergence $-4 < x < 2$, which is the interval of convergence of the series for $f(2x)$?

    (A) $-4 < x < 2$ (B) $-8 < x < 4$ (C) $-2 < x < 1$

    (D) $-2 < x < 4$ (E) Cannot be determined

3. Find the interval of convergence of the series $\sum_{n=0}^{\infty} \frac{(x + 2)^n}{n!}$.

    (A) $-3 < x < -1$

    (B) $1 < x < 3$

    (C) $-5 < x < 1$

    (D) Converges only at $x = -2$

    (E) Converges for all real x

Need More Help With . . .   *See . . .*

        Power series?   *Calculus,* Section 9.1

        Ratio Test?   *Calculus,* Section 9.4

LaGrange Error Bound

AP* Objective:

Taylor series

• LaGrange error bound for Taylor polynomials

Big Picture

When using Taylor polynomials to model functions, error estimation is an important skill. We make use of the LaGrange error bound, sometimes called the Taylor Remainder Estimation Theorem. This challenging topic often appears on the Free Response portion of the AP* Calculus Exam as a portion of the questions on Taylor series.

Content and Practice

When an infinite Taylor series is truncated to form a polynomial of order n, the truncated series only approximates the function. The exact error cannot always be determined, but the following theorem produces an upper bound for the error.

Taylor Remainder Estimation Theorem

If there are positive constants M and r such that $\left|f^{(n+1)}(t)\right| \leq Mr^{n+1}$ for all t between a and x, then the remainder $R_n(x)$ satisfies the inequality

$$\left|R_n(x)\right| \leq M\frac{r^{n+1}\left|x - a\right|^{n+1}}{(n + 1)!}.$$

Understanding the meaning of each term in the formula will remove some of the confusion this formula frequently causes.

• n is the order of the approximating polynomial.

• a is the center of the series.

• x is any value in the interval, I, where the series is approximating f. We choose x to make $\left|x - a\right|^{n+1}$ as large as possible.

• With certain exceptions, r is very often set equal to 1.

- $\left|f^{(n+1)}(t)\right|$ is the absolute value of the $(n + 1)$st derivative of the approximated function. If the $(n + 1)$st derivative is 0, we can use the $(n + 2)$nd derivative.

- Mr^{n+1} is a product that is a least upper bound for $\left|f^{(n+1)}(t)\right|$ on the interval I.

The formula produces an upper bound for the error between the approximating polynomial and the true function value called the LaGrange error bound.

1. A second-order approximation for e^x on the domain $|x| \le 0.5$ is $e^x \approx 1 + x + \dfrac{x^2}{2!}$. Use the Remainder Estimation Theorem to estimate the error. (*Hint: $n = 2$, $a = 0$, and all derivatives of e^x are e^x.*)

2. Write a Taylor polynomial of order 3 centered at $x = 2$ to approximate $f(x) = \dfrac{1}{3 - x}$ on the interval $|x - 2| < 1$. Find the LaGrange error bound.

Additional Practice

1. The hyperbolic sine is defined as $\sinh x = \dfrac{e^x - e^{-x}}{2}$. A third-order Taylor polynomial approximation is $\sinh x \approx x + \dfrac{x^3}{3!}$. If this is used to approximate $\sinh x$ for $|x| \le 2$, which is the LaGrange error bound?

(A) 4.836 (B) 3.627 (C) 2.718 (D) 2.508 (E) 2.418

2. What is the smallest order of Taylor polynomial centered at $x = 1$ which will approximate e^{x-1} on the domain $-1 \le x \le 3$ with LaGrange error bound less than 1?

(A) 3   (B) 5   (C) 7   (D) 9   (E) 11

Need More Help With . . .

LaGrange error bounds?

See . . .

Calculus, Section 9.3

Part IV

Practice Examinations

Calculus AB—Exam 1

Section I, Part A

Time: 55 minutes
Number of questions: 28

NO CALCULATOR MAY BE USED IN THIS PART OF THE EXAMINATION.

Directions: Solve each of the following problems. After examining the form of the choices, decide which is the best of the choices given.

In this test: Unless otherwise specified, the domain of a function f is assumed to be the set of all real numbers x for which $f(x)$ is a real number.

1. What is the x-coordinate of the point of inflection on the graph of $y = \frac{1}{10}x^5 + \frac{1}{2}x^4 - \frac{3}{10}$?

   (A) -4 (B) -3 (C) -1

   (D) $-\frac{3}{10}$ (E) 0

2. The graph of a piecewise-linear function f, for $-3 \le x \le 5$, is shown. What is the value of $\int_{-3}^{5} f(x)\, dx$?

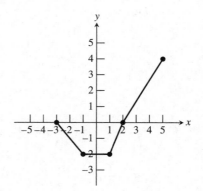

   (A) 16 (B) 13 (C) 4

   (D) 1 (E) -1

3. $\int_2^3 \frac{1}{x^3}\,dx =$

(A) $-\frac{5}{72}$ (B) $-\frac{5}{36}$ (C) $\frac{5}{144}$

(D) $\frac{5}{72}$ (E) $\ln\frac{27}{8}$

4. f is continuous for $a \leq x \leq b$ but not differentiable for some c such that $a < c < b$. Which of the following could be true?

(A) $x = c$ is a vertical asymptote of the graph of f.

(B) $\lim\limits_{x \to c} f(x) \neq f(c)$

(C)   The graph of f has a cusp at $x = c$.

(D) $f(c)$ is undefined.

(E)   None of the above

5. $\int_{\pi/2}^x \cos t\,dt =$

(A) $-\sin x$ (B) $-\sin x - 1$

(C) $\sin x + 1$ (D) $\sin x - 1$

(E) $1 - \sin x$

6. If $x^3 + 2x^2 y - 4y = 7$, then when $x = 1$, $\dfrac{dy}{dx} =$

(A) $-\frac{9}{2}.$ (B)   0.

(C) $-8.$ (D) $-3.$

(E) $\frac{7}{2}.$

7. $\int_1^{e^2} \frac{x^3 + 1}{x}\,dx =$

(A) $\frac{1}{3}e^6 + \frac{8}{3}$ (B) $\frac{1}{3}e^6 + \frac{5}{3}$

(C) $\frac{1}{3}e^6 - \frac{1}{2e^2} + \frac{1}{6}$ (D) $\frac{1}{3}e^6 - \frac{1}{2e^4} + \frac{1}{6}$

(E) $\frac{1}{3}e^6 + \frac{7}{3}$

8. Let f and g be differentiable functions with the following properties:
   I. $f(x) < 0$ for all x
   II. $g(5) = 2$

   If $h(x) = \dfrac{f(x)}{g(x)}$ and $h'(x) = \dfrac{f'(x)}{g(x)}$, then $g(x) =$

   (A) $\dfrac{1}{f'(x)}$.

   (B) $f(x)$.

   (C) $-f(x)$.

   (D) 0.

   (E) 2.

9. The production rate of cola, in thousands of gallons per hour, at the production plant on July 1 is shown in the graph. Of the following, which best approximates the total number of thousands of gallons of cola that were produced that day?

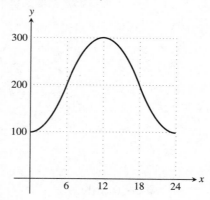

   (A) 800   (B) 4200   (C) 4800

   (D) 5000   (E) 5400

10. What is the instantaneous rate of change at $x = 3$ of the function f given by $f(x) = \dfrac{x^2 - 2}{x + 1}$?

   (A) $-\dfrac{17}{16}$ (B) $-\dfrac{1}{8}$ (C) $\dfrac{1}{8}$

   (D) $\dfrac{13}{16}$ (E) $\dfrac{17}{16}$

11. If f is a linear function and $0 < a < b$, then $\int_b^a f''(x)\,dx =$

    (A) 0   (B) 2   (C) $\dfrac{ab}{2}$

    (D) $m(a - b)$ (E) $\dfrac{a^2 - b^2}{2}$

12. If $f(x) = \begin{cases} \ln 3x, & 0 < x \le 3 \\ x \ln 3, & 3 < x \le 4 \end{cases}$, then $\lim\limits_{x \to 3} f(x)$ is

    (A) $\ln 9$.   (B) $\ln 27$.   (C) $3 \ln 3$.

    (D) $3 + \ln 3$.   (E) nonexistent.

13. The graph of the function f shown in the figure has a horizontal tangent at the point $(-1, -3)$ and a cusp at $(2, 0)$. For what values of x, $-5 < x < 5$, is f not differentiable?

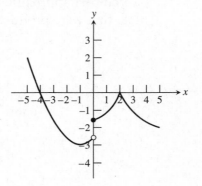

    (A) 0 only   (B) 0 and 2 only   (C) -1 and 0 only

    (D) -1, 0, and 2   (E) -1 and 2 only

14. A particle moves along the x-axis so that its position at time t is given by $x(t) = t^2 - 7t + 12$. For what value of t is the velocity of the particle zero?

    (A) 2.5   (B) 3   (C) 3.5

    (D) 4   (E) 4.5

15. If $F(x) = \int_1^{x^2} \sqrt{t^2 + 3}\, dt$, then $F'(2) =$

   (A) $4\sqrt{19}$ (B) $2\sqrt{19}$ (C) $4\sqrt{7}$

   (D) $2\sqrt{7}$ (E) $\sqrt{7}$

16. If $f(x) = \cos e^{2x}$, then $f'(x) =$

   (A) $\sin e^{2x}$ (B) $2 \sin e^{2x}$ (C) $-\sin e^{2x}$

   (D) $-2 \sin e^{2x}$ (E) $-2e^{2x} \sin e^{2x}$

17. The graph of a twice-differentiable function f is shown. Which of the following is true?

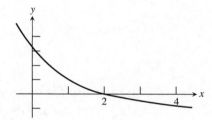

   (A) $f(2) < f'(2) < f''(2)$ (B) $f(2) < f''(2) < f'(2)$

   (C) $f'(2) < f(2) < f''(2)$ (D) $f''(2) < f(2) < f'(2)$

   (E) $f''(2) < f'(2) < f(2)$

18. An equation of the line tangent to the graph of $y = 3x - \cos x$ at $x = 0$ is

   (A) $y = 2x$ (B) $y = 2x - 1$ (C) $y = 3x + 1$

   (D) $y = 3x - 1$ (E) $y = 4x$

19. If $f''(x) = (x - 1)(x + 2)^3(x - 4)^2$, then the graph of f has inflection points when $x =$

(A) -2 only   (B) 1 only   (C) 1 and 4 only

(D) -2 and 1 only   (E) -2, 1, and 4 only

20. What are all values of k for which $\int_{-2}^{k} x^5 \, dx = 0$?

(A) -2 (B) 0   (C) 2

(D) -2 and 2   (E) -2, 0, and 2

21. If $dy/dt = my$ and m is a nonzero constant, then y could be

(A) $4e^{mty}$.   (B) $4e^{mt}$.   (C) $e^{mt} + 4$.

(D) $mty + 4$.   (E) $\frac{m}{2}y^2 + 4$.

22. The function f is given by $f(x) = -x^6 + x^3 - 2$. On which of the following intervals is f decreasing?

(A) $(-\infty, 0)$ (B) $\left(-\infty, -\sqrt[3]{\frac{1}{2}}\right)$ (C) $\left(0, \sqrt[3]{\frac{1}{2}}\right)$

(D) $(0, \infty)$ (E) $\left(\sqrt[3]{\frac{1}{2}}, \infty\right)$

23. The graph of f is shown in the figure below. Which of the following could be the graph of the derivative of f?

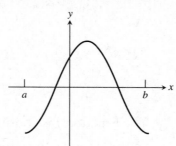

(A)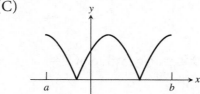

(B)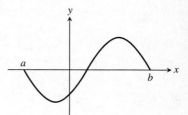

(C)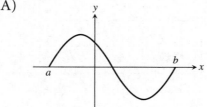

(D)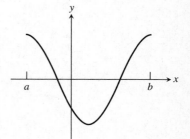

(E)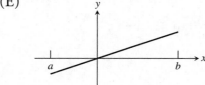

24. The minimum acceleration attained on the interval $0 \leq t \leq 4$ by the particle whose velocity is given by $v(t) = t^3 - 4t^2 - 3t + 2$ is

(A) -16. (B) -10. (C) -8.

(D) $-\dfrac{25}{3}$. (E) -3.

25. What is the area of the region between the graphs of $y = x^3$ and $y = -x - 1$ from $x = 0$ to $x = 2$?

   (A) 0   (B) 4   (C) 5

   (D) 8   (E) 10

26. The function f is continuous on the closed interval $[1, 3]$ and has the values given in the table. The equation $f(x) = \dfrac{5}{4}$ must have at least two solutions in the interval $[1, 3]$ if $k =$

x	1	2	3
$f(x)$	2	k	4

   (A) $\dfrac{1}{4}$.   (B) $\dfrac{3}{2}$.   (C) 2.

   (D) $\dfrac{9}{4}$.   (E) 3.

27. What is the average value of $y = x^3\sqrt{x^4 + 9}$ on the interval $[0, 2]$?

   (A) $\dfrac{98}{3}$ (B) $\dfrac{49}{3}$ (C) $\dfrac{125}{12}$

   (D) $\dfrac{147}{8}$ (E) $\dfrac{49}{6}$

28. If $f(x) = \tan 3x$, then $f'\left(\dfrac{\pi}{9}\right) =$

   (A) $\dfrac{4}{3}$ (B) 4   (C) 6

   (D) 12   (E) $6\sqrt{3}$

Calculus AB—Exam 1
Section I, Part B

Time: 50 Minutes
Number of Questions: 17

A GRAPHING CALCULATOR IS REQUIRED FOR SOME QUESTIONS IN THIS PART OF THE EXAMINATION.

Directions: Solve each of the following problems. After examining the form of the choices, decide which is the best of the choices given.

In this test:

1. The exact numerical value of the correct answer does not always appear among the choices given. When this happens, select from among the choices the number that best approximates the exact numerical value.

2. Unless otherwise specified, the domain of a function f is assumed to be the set of all real numbers x for which $f(x)$ is a real number.

29. The graph of a function f is shown. Which of the following statements about f is false?

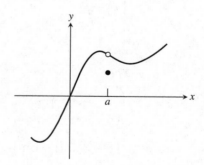

(A) $\lim\limits_{x \to a} f(x)$ exists.

(B) $\lim\limits_{x \to a^-} f(x) = \lim\limits_{x \to a^+} f(x)$

(C) $f(a)$ exists.

(D) f is continuous at $x = a$.

(E) f has a relative minimum at $x = a$.

30. Let $f(x) = 2e^{3x}$ and $g(x) = 5x^3$. At what value of x do the graphs of f and g have parallel tangents?

(A) -0.445 (B) -0.366 (C) -0.344

(D) -0.251 (E) -0.165

31. The side of a cube is increasing at a constant rate of 0.2 centimeter per second. In terms of the surface area S, what is the rate of change of the volume of the cube, in cubic centimeters per second?

(A) $0.1S$ (B) $0.2S$ (C) $0.6S$

(D) $0.04S$ (E) $0.008S$

32. The graphs of the derivatives of the functions f, g, and h are shown. Which of the functions have a relative minimum on the open interval $a < x < b$?

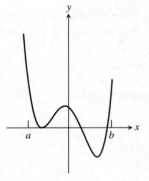

$y = f'(x)$

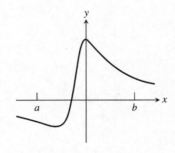

$y = g'(x)$

$y = h'(x)$

(A) f only   (B) g only   (C) h only

(D) f and h only   (E) f, g, and h

33. The first derivative of the function f is given by $f'(x) = \dfrac{\sin^2 x}{x} - \dfrac{2}{9}$. How many critical values does f have on the open interval $(0, 10)$?

(A) One   (B) Two   (C) Three

(D) Four   (E) Six

34. Let f be the function given by $f(x) = x^{2/3}$. Which of the following statements about f are true?

    I. f is continuous at $x = 0$.

    II. f is differentiable at $x = 0$.

    III. f has an absolute minimum at $x = 0$.

(A) I only   (B) II only   (C) III only

(D) I and II only   (E) I and III only

35. If f is a continuous function and if $F'(x) = f(x)$ for all real numbers x, then $\int_{-1}^{2} f(3x)\, dx =$

(A) $3F(2) - 3F(-1)$ (B) $\dfrac{1}{3} F(2) - \dfrac{1}{3} F(-1)$

(C) $F(6) - F(-3)$ (D) $3F(6) - 3F(-3)$

(E) $\dfrac{1}{3} F(6) - \dfrac{1}{3} F(-3)$

36. If $a \neq 0$, then $\lim\limits_{x \to a} \dfrac{x^3 - a^3}{a^6 - x^6}$ is

(A) nonexistent.   (B) 0.   (C) $-\dfrac{1}{2a^3}$.

(D) $-\dfrac{1}{a^3}$.   (E) $\dfrac{1}{2a^3}$.

37. Population P grows according to the equation $dP/dt = kP$, where k is a constant and t is measured in years. If the population doubles every 12 years, then the value of k is approximately

(A) 3.585.   (B) 1.792.   (C) 0.693.

(D) 0.279.   (E) 0.058.

38. The function f is continuous on the closed interval $[1, 9]$ and has the values given in the table. Using the subintervals $[1, 3]$, $[3, 6]$, and $[6, 9]$, what is the value of the trapezoidal approximation of $\int_1^9 f(x)\, dx$?

x	1	3	6	9
$f(x)$	15	25	40	30

(A) 110   (B) 150   (C) 175

(D) 242.5   (E) 262.5

39. The base of a solid is a region in the first quadrant bounded by the x-axis, the y-axis, and the line $x + 3y = 9$, as shown in the figure. If cross sections of the solid perpendicular to the y-axis are isoceles right triangles with the hypotenuses in the xy-plane, what is the volume of the solid?

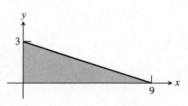

(A) 6.75   (B) 13.5   (C) 15.188

(D) 20.25   (E) 40.5

40. Which of the following is an equation of the line tangent to the graph of $f(x) = x^6 - x^4$ at the point where $f'(x) = -1$?

(A) $y = -x - 1.031$ (B) $y = -x - 0.836$

(C) $y = -x + 0.836$ (D) $y = -x + 0.934$

(E) $y = -x + 1.031$

41. Let $F(x)$ be an antiderivative of $\dfrac{2(\ln x)^4}{3x}$. If $F(2) = 0$, then $F(8) =$

(A) 5.163.   (B) 0.860.   (C) 0.184.

(D) 0.180.   (E) 0.004.

42. If g is a differentiable function such that $g(x) < 0$ for all real numbers x and if $f'(x) = (x^2 - x - 12) g(x)$, which of the following is true?

(A) f has a relative maximum at $x = -3$ and a relative minimum at $x = 4$.

(B) f has a relative minimum at $x = -3$ and a relative maximum at $x = 4$.

(C) f has a relative maximum at $x = 3$ and a relative minimum at $x = -4$.

(D) f has a relative minimum at $x = 3$ and a relative maximum at $x = -4$.

(E) It cannot be determined if f has any relative extrema.

43. If the length l of a rectangle is decreasing at a rate of 2 inches per minute while its width w is increasing at a rate of 2 inches per minute, which of the following must be true about the area A of the rectangle?

(A) A is always increasing.

(B) A is always decreasing.

(C) A is increasing only when $l > w$.

(D) A is increasing only when $l < w$.

(E) A remains constant.

44. Let f be a function that is differentiable on the open interval $(-3, 7)$. If $f(-1) = 4, f(2) = -5$, and $f(6) = 8$, which of the following must be true?

   I. f has at least two zeros.
   II. f has a relative minimum at $x = 2$.
   III. For some c, $2 < c < 6$, $f(c) = 4$.

   (A) I only   (B) II only   (C) I and II only

   (D) I and III only   (E) I, II, and III

45. If $0 \le k \le \dfrac{\pi}{2}$ and the area under the curve $y = \sin x$ from $x = k$ to $x = \dfrac{\pi}{2}$ is 0.75, then $k =$

   (A) 1.318.   (B) 0.848.   (C) 0.723.

   (D) 0.533.   (E) 0.253.

Calculus AB—Exam 1
Section II, Part A

Time: 45 minutes
Number of problems: 3

A GRAPHING CALCULATOR IS REQUIRED FOR SOME PROBLEMS IN THIS PART OF THE EXAMINATION.

1. Let R be the region bounded by the x-axis, the graph of $y = \sqrt{x + 1}$, and the line $x = 3$.

    (A) Find the area of the region R.

    (B) Find the value of h such that the vertical line $x = h$ divides the region R into two regions of equal area.

    (C) Find the volume of the solid generated when R is revolved about the x-axis.

    (D) The vertical line $x = k$ divides the region R into two regions such that when these two regions are revolved about the x-axis, they generate solids with equal volumes. Find the value of k.

2. Let f be the function given by $f(x) = 4xe^{3x}$.

   (A) Find $\lim\limits_{x \to -\infty} f(x)$ and $\lim\limits_{x \to \infty} f(x)$.

   (B) Find the absolute minimum value of f. Justify that your answer is an absolute minimum.

   (C) What is the range of f?

   (D) Consider the family of functions defined by $y = axe^{bx}$, where a and b are nonzero constants, with $a > 0$. Find the absolute minimum value of axe^{bx} in terms of a and c.

3. The temperature outside of a house during a 24-hour period is given by

$$F(t) = 78 - 12 \cos \frac{\pi t}{12}, \quad 0 \le t \le 24,$$

where $F(t)$ is measured in degrees Fahrenheit and t is measured in hours.

   (A) Sketch the graph of F on the grid provided.

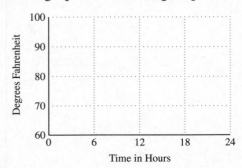

(B) Find the average temperature, to the nearest degree Fahrenheit, between $t = 8$ and $t = 16$.

(C) An air conditioner cooled the house whenever the outside temperature was at or above 76 degrees Fahrenheit. For what values of t was the air conditioner cooling the house?

(D) The cost of cooling the house accumulates at the rate of $0.08 per hour for each degree the outside temperature exceeds 76 degrees Fahrenheit. What was the total cost, to the nearest cent, to cool the house for this 24-hour period?

✧ **End of Part A of Section II** ✧

Calculus AB—Exam 1
Section II, Part B

Time: 45 minutes
Number of problems: 3

NO CALCULATOR MAY BE USED IN THIS PART OF THE EXAMINATION.

4. The graph of the velocity $v(t)$, in feet per second, of a bicycle racing on a straight road, for $0 \leq t \leq 60$, is shown. Also given is a table of values for $v(t)$, at 5 second intervals of time t.

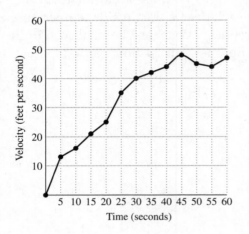

t (seconds)	$v(t)$ (feet per second)
0	0
5	13
10	16
15	21
20	25
25	35
30	40
35	42
40	44
45	48
50	45
55	44
60	47

(A) During what intervals of time is the acceleration of the bike positive? Give a reason for your answer.

(B)   Find the average acceleration of the bike, in feet per second^2, over the interval $0 \le t \le 60$.

(C)   Find one approximation for the acceleration of the bike, in feet per second^2, at $t = 30$. Show the computations you used to arrive at your answer.

(D)   Approximate $\int_0^{60} v(t)\, dt$ with a Riemann sum, using the midpoints of six subintervals of equal length. Using correct units, explain the meaning of this integral.

5.   Consider the curve given by $x^2 + 3y^2 = 1 + 3xy$.

(A)   Show that $\dfrac{dy}{dx} = \dfrac{3y - 2x}{6y - 3x}$.

(B)   Find all points on the curve whose x-coordinate is 1, and write an equation for the tangent line at each of these points.

(C)   Find the coordinates of each point on the curve where the tangent line is vertical.

6. Consider the differential equation $dy/dx = x^2(2y + 1)$.

   (A) On the axes provided, sketch a slope field for the given differential equation at the 12 points indicated.

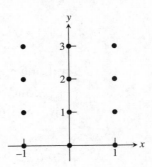

   (B) Although the slope field in part (A) is drawn at only 12 points, it is defined at every point in the xy-plane. Describe all points in the xy-plane for which the slopes are positive.

   (C) Find the particular solution $y = f(x)$ to the given differential equation with the initial condition $f(0) = 5$.

Calculus AB—Exam 2

Section I, Part A

Time: 55 minutes
Number of questions: 28

NO CALCULATOR MAY BE USED IN THIS PART OF THE EXAMINATION.

<u>Directions:</u> Solve each of the following problems. After examining the form of the choices, decide which is the best of the choices given.

<u>In this test:</u> Unless otherwise specified, the domain of a function f is assumed to be the set of all real numbers x for which $f(x)$ is a real number.

1. $\int_1^3 (3x^2 - 4x)\, dx =$

   (A) 8   (B) 9   (C) 10

   (D) 12   (E) 62

2.   If $f(x) = x\sqrt{4x - 1}$, then $f'(x)$ is

   (A) $\dfrac{6x - 1}{\sqrt{4x - 1}}.$ (B) $\dfrac{2x}{\sqrt{4x - 1}}.$ (C) $\dfrac{1}{\sqrt{4x - 1}}.$

   (D) $\dfrac{-6x + 2}{\sqrt{4x - 1}}.$ (E) $\dfrac{9x - 2}{2\sqrt{4x - 1}}.$

3.   If $\int_a^b g(x)\, dx = 4a + b$, then $\int_a^b (g(x) + 7)\, dx =$

   (A) $8b - 11a.$ (B) $8b + 11a.$ (C) $8b - 3a.$

   (D) $7b - 7a.$ (E) $4a + b + 7.$

4. If $f(x) = -x^5 + x + \dfrac{1}{x^2}$, then $f'(-1) =$

    (A) 8.   (B) 2.   (C) −2.

    (D) −3.   (E) −8.

5. $y = 5x^4 - 24x^3 + 24x^2 + 17$ is concave down for

    (A) $x < 0$.   (B) $x > 0$.

    (C) $x < -2$ or $x > -\dfrac{2}{5}$.   (D) $x < \dfrac{2}{5}$ or $x > 2$.

    (E) $\dfrac{2}{5} < x < 2$.

6. $\dfrac{1}{3} \int e^{t/3} \, dt =$

    (A) $e^t + C$ (B) $3e^{t/3} + C$ (C) $e^{t/3} + C$

    (D) $\dfrac{1}{3} e^{t/3} + C$ (E) $e^{-2/3t} + C$

7. $\dfrac{d}{dx} \cos^3 x^2 =$

    (A) $6x \cos^2 x^2$ (B) $\sin^3 x^2$

    (C) $6x \sin x^2 \cos^2 x^2$ (D) $-3 \sin x^2 \cos^2 x^2$

    (E) $-6x \sin x^2 \cos^2 x^2$

Questions 8–9 refer to the following situation.

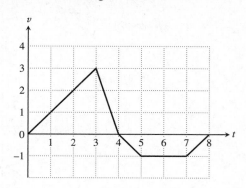

A spider begins to crawl up a vertical blade of grass at time $t = 0$. The velocity v of the spider at time t, $0 \le t \le 8$, is given by the function whose graph is shown.

8. At what value of t does the spider change direction?

    (A) 3 (B) 4 (C) 5

    (D) 7 (E) 8

9. What is the total distance the spider traveled from $t = 0$ to $t = 8$?

    (A) 3 (B) 8 (C) 9

    (D) 10 (E) 15

10. An equation of the line tangent to the graph of $y = \cos 3x$ at $x = \pi/6$ is

    (A) $y = 3\left(x - \dfrac{\pi}{6}\right)$ (B) $y = -\left(x - \dfrac{\pi}{6}\right)$

    (C) $y = -3\left(x - \dfrac{\pi}{6}\right)$ (D) $y - 1 = -\left(x - \dfrac{\pi}{6}\right)$

    (E) $y - 1 = -2\left(x - \dfrac{\pi}{6}\right)$

11. The graph of the derivative of f is shown in the figure below. Which of the following could be the graph of f?

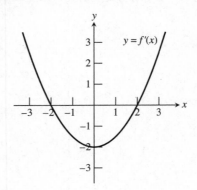

(A)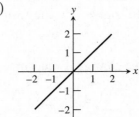

(B)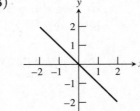

(C)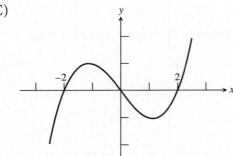

(D)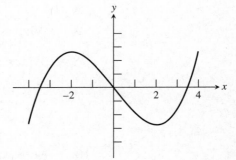

(E)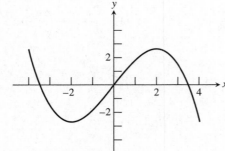

12. At what point on the graph of $y = \frac{1}{2}x^2 - \frac{3}{2}$ is the tangent line parallel to the line $4x - 8y = 5$?

(A) $\left(\frac{1}{2}, -\frac{3}{8}\right)$

(B) $\left(\frac{1}{2}, -\frac{11}{8}\right)$

(C) $\left(2, \frac{3}{8}\right)$

(D) $\left(2, \frac{1}{2}\right)$

(E) $\left(-\frac{1}{2}, -\frac{11}{8}\right)$

13. Let f be a function defined for all real numbers x. If $f'(x) = \dfrac{\left|9 - x^2\right|}{x - 3}$, then f is decreasing on the interval

(A) $(-\infty, 3)$.

(B) $(-\infty, \infty)$.

(C) $(-3, 6)$.

(D) $(-3, \infty)$.

(E) $(3, \infty)$.

14. Let f be a differentiable function such that $f(5) = 3$ and $f'(5) = 2$. If the tangent line to the graph of f at $x = 5$ is used to find an approximation to a zero of f, that approximation is

(A) 6.5.

(B) 4.3.

(C) 3.5.

(D) 0.5.

(E) 0.3.

15. The graph of the function f is shown. Which of the following statements about f is true?

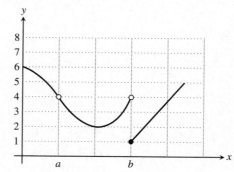

(A) $\lim\limits_{x \to a} f(x) = \lim\limits_{x \to b} f(x)$

(B) $\lim\limits_{x \to a} f(x) = 4$

(C) $\lim\limits_{x \to b} f(x) = 4$

(D) $\lim\limits_{x \to b} f(x) = 1$

(E) $\lim\limits_{x \to a} f(x)$ does not exist.

16. The area of the region enclosed by the graph of $y = x^2 + 2$ and the line $y = 11$ is

(A) 18.

(B) 30.

(C) 36.

(D) 72.

(E) 27π.

17. If $x^2 = 25 - y^2$, what is the value of $\dfrac{d^2y}{dx^2}$ at the point $(3, 4)$?

(A) $\dfrac{-25}{64}$

(B) $\dfrac{-7}{64}$

(C) $\dfrac{7}{64}$

(D) $\dfrac{25}{64}$

(E) $\dfrac{4}{3}$

18. $\displaystyle\int_{\pi/4}^{\pi/2} \dfrac{-e^{\cot x}}{\sin^2 x}\, dx =$

(A) $-e$

(B) $1 - e$

(C) -1

(D) $e - 1$

(E) $1 + e$

19. If $f(x) = \ln|1 - x^2|$, then $f'(x) =$

(A) $\dfrac{-2|x|}{1 - x^2}.$

(B) $\dfrac{-2x}{1 - x^2}.$

(C) $\dfrac{1}{1 - x^2}.$

(D) $\left|\dfrac{-2x}{1 - x^2}\right|.$

(E) $\dfrac{-2x}{|1 - x^2|}.$

20. The average value of $f(x) = -\sin x$ on the interval $[-2, 4]$ is

(A) $\dfrac{\cos 4 + \cos 2}{6}.$

(B) $\dfrac{\cos 2 - \cos 4}{2}.$

(C) $\dfrac{\cos 4 + \cos 2}{2}.$

(D) $\dfrac{\cos 4 - \cos 2}{2}.$

(E) $\dfrac{\cos 4 - \cos 2}{6}.$

21. Evaluate $\lim\limits_{x \to 1} \dfrac{\ln x}{3x}$

    (A) 0   (B) $\dfrac{3}{e}$ (C) e

    (D) 3   (E) The limit does not exist.

22. What are all values of x for which the function f defined by $f(x) = (x^2 - 15)e^{-x}$ is increasing?

    (A) There are no such values of x.
    (B) $x < -3$ or $x > 5$
    (C) $-5 < x < 3$
    (D) $-3 < x < 5$
    (E) All values of x

23. If the region enclosed by the y-axis, the line $y = 2$, and the curve $y = \sqrt[3]{x}$ is revolved about the y-axis, the volume of the solid generated is

    (A) π.   (B) 4π.   (C) 8π.

    (D) $\dfrac{64\pi}{7}$.   (E) $\dfrac{128\pi}{7}$.

24. The expression $\dfrac{1}{30}\left(\sin\dfrac{1}{30} + \sin\dfrac{2}{30} + \sin\dfrac{3}{30} + \cdots + \sin\dfrac{30}{30}\right)$ is a Riemann sum approximation for

    (A) $\int_0^1 \sin\dfrac{x}{30}\, dx$.   (B) $\int_0^1 \sin x \, dx$.

    (C) $\dfrac{1}{30}\int_0^1 \sin\dfrac{x}{30}\, dx$.   (D) $\dfrac{1}{30}\int_0^1 \sin x \, dx$.

    (E) $\dfrac{1}{30}\int_0^{30} \sin x \, dx$.

25. $\int x \sin x^2 \, dx =$

    (A) $-\dfrac{1}{2}\cos x^2 + C$ (B) $\dfrac{1}{2}\cos x^2 + C$

    (C) $-x^2 \cos x^2 + C$ (D) $x^2 \cos x^2 + C$

    (E) $\dfrac{1}{2}x^2 \cos\dfrac{x^2}{3} + C$

26. Let $f(x) = \lim\limits_{h \to 0} \dfrac{(x + h)^2 - x^2}{h}$. For what value of x does $f(x) = 4$?

(A) -2 (B) -1 (C) 1

(D) 2   (E) 4

27. Let $f(x) = \begin{cases} 3x^2 - 5, & x \le 1 \\ 6x + 2, & x > 1 \end{cases}$. Which of the following are true statements about this function?

    I. $\lim\limits_{x \to 1} f(x)$ exists.

    II. $\lim\limits_{x \to 1} f'(x)$ exists.

    III. $f'(1)$ exists.

(A) None   (B) II only   (C) III only

(D) II and III   (E) I, II, and III

28. Let $g(x) = \dfrac{d}{dx} \int_0^x \sqrt{t^2 + 9}\, dt$. What is $g(-4)$?

(A) -5 (B) -3 (C) 3

(D) 4   (E) 5

⬧ **End of Part A of Section I** ⬧

Calculus AB—Exam 2
Section I, Part B

Time: 50 minutes
Number of questions: 17

A GRAPHING CALCULATOR IS REQUIRED FOR SOME QUESTIONS IN THIS PART OF THE EXAMINATION.

<u>Directions:</u> Solve each of the following problems. After examining the form of the choices, decide which is the best of the choices given.

<u>In this test:</u>

1. The exact numerical value of the correct answer does not always appear among the choices given. When this happens, select from among the choices the number that best approximates the exact numerical value.

2. Unless otherwise specified, the domain of a function f is assumed to be the set of all real numbers x for which $f(x)$ is a real number.

29. If $f(x) = \dfrac{e^{3x}}{3x}$, then $f'(x) =$

   (A) 1.

   (B) e^{3x}.

   (C) $\dfrac{e^{3x}(1 - 3x)}{3x^2}$.

   (D) $\dfrac{e^{3x}(3x + 1)}{3x^2}$.

   (E) $\dfrac{e^{3x}(3x - 1)}{3x^2}$.

30. The graph of the function $y = \dfrac{1}{3}x^3 - x^2 - 5x + 3 \sin x$ changes concavity at $x =$

   (A) 3.29.

   (B) 2.21.

   (C) 1.34.

   (D) 0.41.

   (E) -0.39.

31. The graph of f is shown. If $\int_1^4 f(x)\,dx = 3.8$ and $F'(x) = f(x)$, then $F(4) - F(0) =$

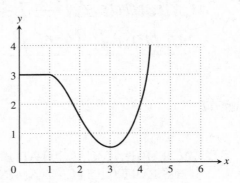

(A) 0.8.   (B) 2.8.   (C) 4.8.

(D) 6.8.   (E) 8.4.

32. Let f be a function such that $\lim\limits_{h\to 0}\dfrac{f(7+h)-f(7)}{h} = 4$. Which of the following must be true?

   I. f is continuous at $x = 7$.

   II. f is differentiable at $x = 7$.

   III. The derivative of f is continuous at $x = 7$.

(A) I only   (B) II only   (C) I and II only

(D) I and III only   (E) II and III only

33. Let f be the function given by $f(x) = 5e^{3x^3}$. For what positive value of a is the slope of the line tangent to the graph of f at $(a, f(a))$ equal to 6?

(A) 0.142   (B) 0.344   (C) 0.393

(D) 0.595   (E) 0.714

34. Two roads cross at right angles, one running north/south and the other east/west. Eighty feet south of the intersection is an old radio tower. A car traveling at 50 feet per second passes through the intersection heading east. At how many feet per second is the car moving away from the radio tower 3 seconds after it passes through the intersection?

(A) 43.65   (B) 44.12   (C) 44.59

(D) 56.67   (E) 81.76

35. If $y = 3x + 6$, what is the minimum value of $x^3 y$?

(A) -10.125 (B) -5.0625 (C) -1.5

(D) 0   (E) 1.5

36. What is the area of the region in the first quadrant enclosed by the graphs of $y = \sin x$, $y = 2 - x$, and the x-axis?

(A) 0.552   (B) 0.951   (C) 1.106

(D) 1.600   (E) 2.152

37. The base of a solid S is the region enclosed by the graph of $y = \sqrt{\ln(x - 1)}$, the line $x = 2e$, and the x-axis. If the cross sections of S perpendicular to the x-axis are squares, then the volume of S is

(A) 1.587.   (B) 2.173.   (C) 3.185.

(D) 3.501.   (E) 6.347.

38. If the derivative of f is given by $f'(x) = 2e^x - 5x^2$, at which of the following values of x does f have a relative maximum value?

(A) -0.494 (B) 0.259   (C) 1.092

(D) 2.543   (E) 3.310

39. Let $f(x) = \sqrt{2x}$. If the rate of change of f at $x = c$ is four times its rate of change at $x = 1$, then $c =$

(A) $\dfrac{1}{16}$.

(B) $\dfrac{1}{2\sqrt{2}}$.

(C) $\dfrac{1}{\sqrt{2}}$.

(D) 1.

(E) 32.

40. At time $t \geq 0$, the acceleration of a particle that is moving along the x-axis is $a(t) = t + 2 \sin t$. At $t = 0$, the velocity of the particle is -4. For what value of t will the velocity of the particle be zero?

(A) 0

(B) 1.20

(C) 1.78

(D) 2.31

(E) 3.87

41. Let $f(x) = \int_a^x h(t)\, dt$, where h has the graph shown below. Which of the following could be the graph of f?

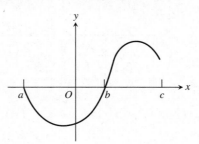

(A)

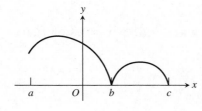

(B)

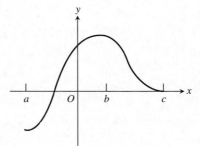

(C)

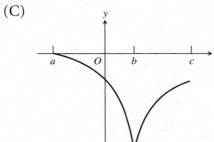

(D)

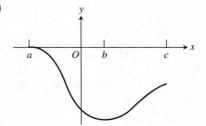

(E)

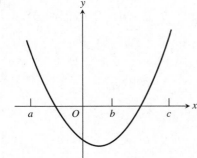

42. A continuous function $f(x)$ has the values shown in the table. What is the value of a trapezoidal approximation of $\int_0^3 f(x)\, dx$ using six equal subintervals?

x	0	0.5	1.0	1.5	2.0	2.5	3.0
$f(x)$	8	5	4	3	3	5	8

(A) 9   (B) 14   (C) 18

(D) 28   (E) 56

43. Which of the following are antiderivatives of $f(x) = 4 \sin x \cos x$?

    I. $F(x) = -\cos 2x$

   II. $F(x) = 2 \sin^2 x$

  III. $F(x) = -2 \cos^2 x$

(A) I only   (B) II only   (C) III only

(D) I and II   (E) I, II, and III

44. Let f be a function such that $f''(x) < 0$ for all x in the closed interval $[3, 4]$, with selected values shown in the table. Which of the following must be true about $f'(3.3)$?

x	3.2	3.3	3.4	3.5
$f(x)$	2.48	2.68	2.86	3.03

(A) $f'(3.3) < 0$ (B) $0 < f'(3.3) < 1.6$

(C) $1.6 < f'(3.3) < 1.8$ (D) $1.8 < f'(3.3) < 2.0$

(E) $f'(3.3) > 2.0$

45. If the function f is defined by $f(x) = \int_0^x -\sin t^2 \, dt$ on the closed interval $-1 \leq x \leq 3$, then f has a local maximum at $x =$

(A) -1.084. (B) 0. (C) 1.772.

(D) 2.171. (E) 2.507.

Calculus AB—Exam 2
Section II, Part A

Time: 45 minutes
Number of problems: 3

A GRAPHING CALCULATOR IS REQUIRED FOR SOME PROBLEMS IN THIS PART OF THE EXAMINATION.

1. A particle moves along the x-axis in such a way that its velocity at any time $t \geq 0$ is given by $v(t) = 3t^2 - 4t - 4$. The particle's position $x(t)$ has a value of 1 when $t = 1$.

    (A) Write a polynomial expression for the position of the particle at any time $t \geq 0$.

    (B) For what value(s) of t, $0 \leq t \leq 4$, is the particle's instantaneous velocity the same as its average velocity on the closed interval $[0, 4]$?

    (C) Find the total distance traveled by the particle from time $t = 0$ until time $t = 4$.

2. Let f be the function given by $f(x) = 4 \sin x$. As shown, the graph of f passes through the point $M(\pi/2, 4)$ and crosses the x-axis at point $N(\pi, 0)$.

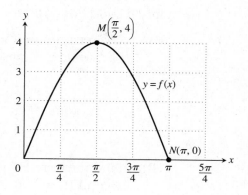

(A) Write an equation for the line passing through points M and N.

(B) Write an equation for the line tangent to the graph of f at point N. Show the analysis that leads to your equation.

(C) Find the x-coordinate of the point on the graph of f, between points M and N, at which the line tangent to the graph of f is parallel to line MN.

(D) Let R be the region in the first quadrant bounded by the graph of f and line segment MN. Write an integral expression for the volume of the solid generated by revolving the region R about the x-axis. DO NOT EVALUATE.

3.	A crate of supplies is dropped from an airplane with a remote-controlled parachute. Let $v(t)$ be the velocity (in meters per second) of the crate at time t seconds, $t \geq 0$. After the parachute opens, the velocity of the crate satisfies the differential equation $dv/dt = -3v - 15$, with initial condition $v(0) = -15$.

(A)	Use separation of variables to find an expression for v in terms of t, where t is measured in seconds.

(B)	Find the terminal velocity of the crate to the nearest meter per second [terminal velocity is defined as $\lim_{t \to \infty} v(t)$].

(C)	It is safe for the package to land once the crate is dropping no more than 5.3 meters per second. At what time t does the crate reach this speed?

Calculus AB—Exam 2
Section II, Part B

Time: 45 minutes
Number of problems: 3

NO CALCULATOR MAY BE USED IN THIS PART OF THE EXAMINATION.

4.   Let f be the function given by $f(x) = 3\sqrt{x-2}$.

   (A)   On the axes provided, sketch the graph of f and shade the region R enclosed by the graph of f, the x-axis, and the vertical line $x = 8$.

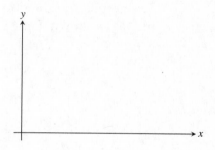

   (B)   Find the area of the region R described in part (A).

   (C)   Rather than using the line $x = 8$ as in part (A), consider the line $x = k$, where k can be any number greater than 2. Let $A(k)$ be the area of the region enclosed by the graphs of f, the x-axis, and the vertical line $x = k$. Write an integral expression for $A(k)$.

   (D)   Let $A(k)$ be as described in part (C). Find the rate of change of A with respect to k when $k = 8$.

5. Let f be the function given by $f(x) = x^3 - 3x^2 + k$, where k is an arbitrary constant.
    (A) Write an expression for $f'(x)$ and use it to find the relative maximum and minimum values for f in terms of k. Show the analysis that leads to your conclusion.

    (B) For what values of the constant k does f have three distinct real zeros?

    (C) Find the value of k such that the average value of f over the closed interval $[-2, 1]$ is 2.

6. The graph of a function f consists of a semicircle and two line segments, as shown. Let g be the function given by $g(x) = \int_0^x f(t)\, dt$.

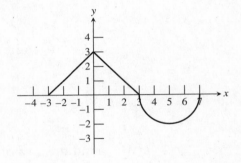

    (A) Find $g(5)$.

    (B) Find all values of x on the open interval $(-3, 7)$ at which g has a relative maximum. Justify your answer.

    (C) Write an equation for the line tangent to the graph of g at $x = 5$.

    (D) Find the x-coordinate of each point of inflection of the graph of g on the open interval $(-3, 7)$. Justify your answer.

Calculus BC—Exam 1

Section I, Part A

Time: 55 minutes
Number of questions: 28

NO CALCULATOR MAY BE USED IN THIS PART OF THE EXAMINATION.

<u>Directions:</u> Solve each of the following problems. After examining the form of the choices, decide which is the best of the choices given.

<u>In this test:</u> Unless otherwise specified, the domain of a function f is assumed to be the set of all real numbers x for which $f(x)$ is a real number.

1. $\int_0^1 \sqrt{x}\,(x^2 + 1)\,dx =$

   (A) $\dfrac{4}{3}$ (B) $\dfrac{9}{7}$ (C) $\dfrac{16}{15}$

   (D) $\dfrac{20}{21}$ (E) $\dfrac{4}{21}$

2. If $x = e^{4t}$ and $y = \sin 6t$, then $\dfrac{dy}{dx} =$

   (A) $\dfrac{3e^{-4t}\cos 6t}{2}.$ (B) $-\dfrac{3\cos 6t}{2e^{4t}}.$ (C) $\dfrac{3e^{-4t}\cos t}{2}.$

   (D) $e^{-4t}\cos 6t.$ (E) $6\cos 6t.$

3. The function f defined by $f(x) = x^4 - x^2$ has a relative minimum at $x =$

   (A) $\sqrt{2}.$ (B) $1.$ (C) $\dfrac{\sqrt{2}}{2}.$

   (D) $\dfrac{1}{2}.$ (E) $0.$

4. $\dfrac{d}{dx} x^2 e^{\ln x^3} =$

(A) $6x^3$ (B) $5x^4$ (C) $2x + 3x^2$

(D) $2x^4 + x^5$ (E) $6x^4$

5. If $g(x) = \dfrac{1}{4} e^{2x-6} + (x - 2)^{5/2}$, then $g'(3) =$

(A) 3.   (B) $\dfrac{5}{2}$.   (C) $\dfrac{11}{4}$.

(D) $\dfrac{1}{4}$.   (E) 0.

6. Find the slope of the line normal to the curve $y = -\sqrt{x + 4}$ at the point where $x = 0$.

(A) -4 (B) $-\dfrac{1}{4}$ (C) $-\dfrac{1}{8}$

(D) $\dfrac{1}{4}$ (E) 4

7. Compute $\dfrac{dy}{dx}$ for the relation $2xy^2 + 3 \ln y = x^2 - 3y^3$ at the point $(3, 1)$.

(A) $\dfrac{5}{2}$ (B) $\dfrac{3}{4}$ (C) $\dfrac{5}{12}$

(D) $\dfrac{1}{8}$ (E) $\dfrac{1}{6}$

8. $\displaystyle\int_1^\infty \dfrac{x^2}{(1 + x^3)^2}\, dx$ is

(A) $-\dfrac{1}{6}$.   (B) $-\dfrac{1}{24}$.   (C) $\dfrac{1}{24}$.

(D) $\dfrac{1}{6}$.   (E) divergent.

The function $h(x)$ is continuous and differentiable on the domain $[0, 7]$. The graph of $h'(x)$ is shown. Use the graph for Questions 9, 10, and 11.

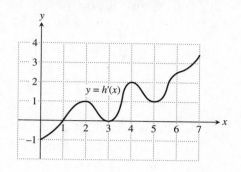

9. At what value of x does $h(x)$ have its absolute minimum?

    (A) 0   (B) 1   (C) 3

    (D) 5   (E) 7

10. The point $(5, 2)$ is on the graph of $y = h(x)$. An equation of the line tangent to $h(x)$ at $(5, 2)$ is

    (A) $y - 2 = x - 5$.   (B) $y = x - 2$.

    (C) $y - 2 = 2(x - 5)$.   (D) $x = 5$.

    (E) $y = 2$.

11. How many inflection points does h have on the interval $(0, 7)$?

    (A) 3   (B) 4   (C) 5

    (D) 6   (E) 7

12. The sum of the infinite geometric series
    $$\frac{8}{25} - \frac{24}{125} + \frac{72}{625} - \frac{216}{3125} + \cdots \text{ is}$$

    (A) 0.2.   (B) 0.6.   (C) 0.8.

    (D) 1.0.   (E) 1.2.

13. A particle moves along the x-axis so that its acceleration at any time t is $a(t) = 2t - 3$. If the initial velocity of the particle is -4, at what time t in the time interval $0 \le t \le 5$ is the particle farthest left?

(A) 0   (B) $\frac{3}{2}$ (C) 3

(D) 4   (E) 5

14. The graph of $f'(x)$ is shown. It is tangent to the x-axis at point c. Which of the following describes all relative extrema of $f(x)$ on the open interval (a, b)?

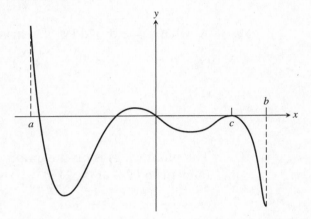

(A) One relative maximum and one relative minimum
(B) One relative maximum and two relative minima
(C) Three relative maxima and two relative minima
(D) Two relative maxima and two relative minima
(E) Two relative maxima and one relative minimum

15. The length of the path described by the parametric equations $x = 2 \sin t$ and $y = 3 \cos t$ for $0 \le t \le \pi/2$ is given by

(A) $\int_0^{\pi/2} \sqrt{4 + 5 \sin^2 t} \, dt$.

(B) $\int_0^{\pi/2} \sqrt{4 \cos^2 t - 9 \sin^2 t} \, dt$.

(C) $\int_0^{\pi/2} \sqrt{4 \sin^2 t + 9 \cos^2 t} \, dt$.

(D) $\int_0^{\pi/2} \sqrt{1 + \dfrac{9 \sin^2 t}{4 \cos^2 t}} \, dt$.

(E) none of the above.

16. $\lim\limits_{x \to 3} \dfrac{e^{x^2} - e^9}{x - 3} =$

    (A) 0   (B) $\dfrac{e^9}{3}$ (C) $3e^9$

    (D) $6e^9$ (E) ∞

17. Let f be the function defined by $f(x) = \ln(x + 4)$. The third-degree Taylor polynomial for f centered about $x = -3$ is

    (A) $x - 3 - \dfrac{(x - 3)^2}{2} - \dfrac{(x - 3)^3}{3}$.

    (B) $x - 3 - \dfrac{(x - 3)^2}{2} + \dfrac{(x - 3)^3}{3}$.

    (C) $-3 + x - \dfrac{(x + 3)^2}{2} + \dfrac{(x + 3)^3}{3}$.

    (D) $3 + x + \dfrac{(x + 3)^2}{2} + \dfrac{(x + 3)^3}{3}$.

    (E) $3 + x - \dfrac{(x + 3)^2}{2} + \dfrac{(x + 3)^3}{3}$.

18. For what values of t does the curve defined by the parametric equations $x = \frac{4}{3} t^3 - t^2$ and $y = t^5 + t^2 - 7t$ have a vertical tangent?

    (A) 0 only   (B) 0 and $\frac{1}{2}$ (C) $\frac{1}{2}$ only

    (D) 1 only   (E) $0, \frac{1}{2}$, and 1

19. The graph of $y = f(x)$ is shown. Let A and B be positive numbers that represent the area of each shaded region. Evaluate $\int_3^{-1} f(x)\, dx + 3\int_{-1}^2 f(x)\, dx$ in terms of A and B.

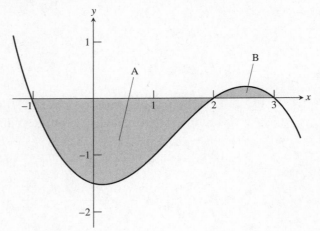

(A) $-2A - B$ (B) $2A + B$ (C) $3A - B$

(D) $3A + B$ (E) $-3A - B$

20. What are all values of x for which the series $\displaystyle\sum_{n=1}^{\infty} \frac{(2x-1)^n}{n \cdot 4^n}$ converges?

(A) $-\dfrac{3}{2} \le x \le \dfrac{5}{2}$ (B) $-\dfrac{3}{2} \le x < \dfrac{5}{2}$

(C) $-\dfrac{3}{2} < x \le \dfrac{5}{2}$ (D) $-3 \le x < 5$

(E) $-3 < x \le 5$

21. The expression representing the area inside one leaf of the polar rose $r = 3\cos 2\theta$ is given by

(A) $\int_0^{\pi/4} \sqrt{1 + 9\sin^2 2\theta}\, d\theta$ (B) $\int_0^{\pi/2} \sqrt{1 + 9\sin^2 2\theta}\, d\theta$

(C) $\int_0^{\pi/2} 9\cos^2 2\theta\, d\theta$ (D) $\dfrac{1}{2}\int_0^{\pi/4} 9\cos^2 2\theta\, d\theta$

(E) $\int_0^{\pi/4} 9\cos^2 2\theta\, d\theta$

22.

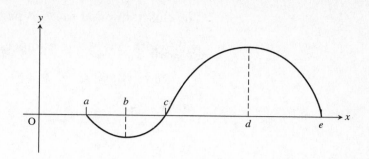

The graph of f is shown. If $h(x) = \int_a^x f(t)\, dt$, for what value of x does $h(x)$ have its minimum?

(A) a (B) b (C) c

(D) d (E) e

23. In the right triangle shown, θ is increasing at a constant rate of 2 radians per minute. In units per minute, at what rate is x increasing when $x = 12$?

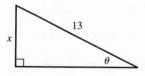

(A) 2   (B) 4   (C) 5

(D) 10   (E) 24

24. The Taylor series for $\cos x$ about $x = 0$ is $1 - \dfrac{x^2}{2!} + \dfrac{x^4}{4!} - \dfrac{x^6}{6!} + \cdots$. If h is a function such that $h'(x) = \cos x^3$, then the coefficient of x^7 in the Taylor series for $h(x)$ about $x = 0$ is

(A) $-\dfrac{1}{14}$.   (B) $-\dfrac{1}{7!}$.   (C) 0.

(D) $\dfrac{1}{7!}$.   (E) $\dfrac{1}{14}$.

25. The closed interval $[w, v]$ is partitioned into k equal subintervals, each with width Δx, by the numbers $x_0, x_1, \ldots, x_k$ with $w = x_0 < x_1 < x_2 < \cdots < x_{k-1} < x_k = v$. The $\displaystyle \lim_{k \to \infty} \sum_{j=1}^{k} \frac{1}{\sqrt{x_j}} \Delta x$ equals

(A) $\sqrt{v} - \sqrt{w}.$

(B) $2\left(\sqrt{v} - \sqrt{w}\right).$

(C) $\dfrac{1}{\sqrt{v}} - \dfrac{1}{\sqrt{w}}.$

(D) $2\left(\dfrac{1}{\sqrt{v}} - \dfrac{1}{\sqrt{w}}\right).$

(E) $\dfrac{2}{3}\left(v^{3/2} - w^{3/2}\right).$

26. $\displaystyle \int \frac{6x - 8}{(x - 3)(x + 2)}\, dx =$

(A) $2 \ln |x - 3| + 4 \ln |x + 2| + C$

(B) $2 \ln |x - 3| + 2 \ln |x + 2| + C$

(C) $2 \ln |x + 3| + 4 \ln |x - 2| + C$

(D) $6x \ln |x - 3| + 8 \ln |x + 2| + C$

(E) $4 \ln |x - 3| + 2 \ln |x + 2| + C$

27. $\displaystyle \int 2x \cos x\, dx =$

(A) $x^2 \sin x + C$

(B) $x^2 \cos \dfrac{x^2}{2} + C$

(C) $2 \sin x - 2x \cos x + C$

(D) $-2x \sin x - 2 \cos x + C$

(E) $2x \sin x + 2 \cos x + C$

28.   Which of the following equations has the slope field shown?

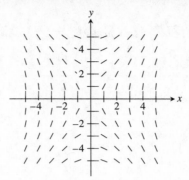

(A) $\dfrac{dy}{dx} = 2x$ (B) $\dfrac{dy}{dx} = 2y$ (C) $\dfrac{dy}{dx} = \dfrac{2x}{y}$

(D) $\dfrac{dy}{dx} = xy$ (E) $\dfrac{dy}{dx} = \dfrac{2y}{x}$

Calculus BC—Exam 1
Section I, Part B

Time: 50 Minutes
Number of Questions: 17

A GRAPHING CALCULATOR IS REQUIRED FOR SOME QUESTIONS IN THIS PART OF THE EXAMINATION.

<u>Directions:</u> Solve each of the following problems. After examining the form of the choices, decide which is the best of the choices given.

<u>In this test:</u>

1. The exact numerical value of the correct answer does not always appear among the choices given. When this happens, select from among the choices the number that best approximates the exact numerical value.

2. Unless otherwise specified, the domain of a function f is assumed to be the set of all real numbers x for which $f(x)$ is a real number.

29. Which of the following sequences converge?

   I. $\left\{ \dfrac{2^n}{n+1} \right\}$

   II. $\left\{ \dfrac{3n+7}{8n} \right\}$

   III. $\left\{ \dfrac{e^{2n}}{e^n + 3^n} \right\}$

   (A) I only   (B) II only   (C) I and II only

   (D) II and III only   (E) None of them

30. When the region enclosed by the graphs of $y = 2x$ and $y = 6x - x^2$ is revolved around the y-axis, the volume of the solid generated is given by

(A) $\pi \int_0^4 \left(8x^2 - 2x^3\right) dx.$ (B) $2\pi \int_0^{16} x\left(4x - x^2\right) dx.$

(C) $2\pi \int_0^4 \left(4x - x^2\right) dx.$ (D) $\pi \int_0^4 \left[\left(6x - x^2\right)^2 - \left(2x\right)^2\right] dx.$

(E) $2\pi \int_0^4 x\left(x^2 - 4x\right) dx.$

31. $\displaystyle\lim_{h \to 0} \frac{\ln\left(\frac{1}{e} + h\right) + 1}{h}$ is

(A) $f'(e)$ where $f(x) = \frac{1}{x}$.

(B) $f'(e)$ where $f(x) = -\ln\frac{1}{x}$.

(C) $f'(1)$ where $f(x) = \ln\frac{x}{e}$.

(D) $f'\left(\frac{1}{e}\right)$ where $f(x) = \ln(x + e)$.

(E) $f'\left(\frac{1}{e}\right)$ where $f(x) = \ln x$.

32. The position of an object oscillating on the x-axis is given by $x(t) = 2\sin 4t - \cos 4t$, where t is the time in seconds. In the first 5 seconds, how many times is the velocity of the object equal to 0?

(A) 0   (B) 4   (C) 5

(D) 6   (E) 7

33. Let h be the function defined by $h(x) = \cos 3x + \ln 4x$. What is the least value of x at which the graph of h changes concavity?

(A) 1.555   (B) 0.621   (C) 0.371

(D) 0.096   (E) 0.004

34. Let f be a continuous function on the closed interval $[-2, 5]$. If $f(-2) = 3$ and $f(5) = -7$, then the Intermediate Value Theorem guarantees that

(A) $-7 \le f(x) \le 3$ for all x between -2 and 5.

(B) $f'(c) = -\dfrac{10}{7}$ for at least one c between -2 and 5.

(C) $f(c) = -3$ for at least one c between -2 and 5.

(D) $f(c) = 0$ for at least one c between -7 and 3.

(E) $f(x)$ is continually decreasing between -2 and 5.

35. Which of the following is the closest to the lowest value of x in the interval $0 \le x \le 6$ such that $\int_0^x (t^2 - 3t)\, dt \le \int_2^x t\, dt$?

(A) 0 (B) 1.1075 (C) 1.663

(D) 1.745 (E) 5.823

36. If $\dfrac{dy}{dx} = (\ln x + 2)y$ and $y = 2$ when $x = 1$, then $y =$

(A) $2e^{x \ln x}$. (B) $e^{x + x \ln x}$. (C) $e^{2 + x \ln x}$.

(D) $2e^{(1/x + 2x - 3)}$. (E) $\sqrt{4x + 4 + 2x \ln x}$.

37. The isosceles trapezoid shown has legs and the top base of length r. The acute angle θ that maximizes the area of the trapezoid is

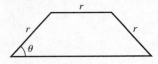

(A) 15°. (B) 30°. (C) 45°.

(D) 60°. (E) 75°.

38. Let f be a twice-differentiable function such that $f(2) = 8$ and $f(4) = 5$. Which of the following must be true for the function f on the interval $2 \le x \le 4$?

   I.   The average value of f is $\dfrac{13}{2}$.

   II. The average rate of change of f is $-\dfrac{3}{2}$.

   III. The average value of f' is $-\dfrac{3}{2}$.

   (A) II only   (B) III only   (C) I and II only

   (D) II and III only (E) I, II, and III

39. Find all values c that satisfy the Mean Value Theorem for the function $f(x) = \dfrac{2}{(1 + x^2)}$ on the interval $[-1, 2]$.

   (A) 0.050   (B) -0.050

   (C) 0.102 and 1.801   (D) 0.050 and 2.449

   (E) None exist in the interval.

40. The base of a solid is the region in the first quadrant enclosed by the graph of $y = 9 - x^2$ and the coordinate axes. If every cross section of the solid perpendicular to the y-axis is a base of a rectangle and each height is three times the base, the volume of the solid is given by

   (A) $\int_0^9 (9 - y)\, dy$.   (B) $3\int_0^9 (9 - x^2)^2\, dx$.

   (C) $3\int_0^9 (9 - y)\, dy$.   (D) $3\int_0^3 (9 - x^2)^2\, dx$.

   (E) $3\int_0^9 (9 + y)\, dy$.

41. Let $f(x) = \int_0^{2x^2} \cos t\, dt$. At how many points in the closed interval $\left[0, \sqrt{\pi}\right]$ does the instantaneous rate of change of f equal the average rate of change of f on that interval?

   (A) 1   (B) 2   (C) 3

   (D) 4   (E) 5

42. If h is an antiderivative of $g(x) = \dfrac{x^3}{1 + x^5}$ and $h(1) = 2$, then $h(3) =$

(A) 4.407.

(B) 2.555.

(C) 1.852.

(D) 0.555.

(E) -0.703.

43. A force of 12 pounds stretches a spring 3 inches beyond its natural length. Assuming Hooke's Law applies, how much work is done stretching the spring from its natural length to 6 inches beyond its natural length?

(A) 4 inch-pounds

(B) 8 inch-pounds

(C) 36 inch-pounds

(D) 54 inch-pounds

(E) 72 inch-pounds

44. The length of the polar curve $r = 1 - 2 \cos 2\theta$ is

(A) 6.925.

(B) 10.008.

(C) 13.365.

(D) 17.629.

(E) 20.016.

45. If $\dfrac{dy}{dx} = x + y$ and $f(0) = 2$, use Euler's method with $\Delta x = 0.5$ to approximate $f(1)$.

(A) 1.75

(B) 3

(C) 4

(D) 4.75

(E) 5.75

Calculus BC—Exam 1
Section II, Part A

Time: 45 minutes
Number of problems: 3

A GRAPHING CALCULATOR IS REQUIRED FOR SOME PROBLEMS IN THIS PART OF THE EXAMINATION.

1. A particle moving along a curve in the plane has position $(x(t), y(t))$ at time t, where

$$\frac{dx}{dt} = \sqrt{t^2 + 4} \qquad \text{and} \qquad \frac{dy}{dt} = 3e^t + 2e^{-t}$$

for all real values of t. At time $t = 0$, the position of the particle is $(3, 4)$.

   (A) Find the speed and acceleration vector of the particle at time $t = 0$.

   (B) Find the equation of the line tangent to the path of the particle at time $t = 0$.

   (C) Find the total distance traveled by the particle over the time interval $0 \le t \le 2$.

   (D) Find the x-coordinate of the position of the particle at time $t = 2$.

2. Let $P(x) = 8 - 4(x - 3) + 5(x - 3)^2 - 7(x - 3)^3 + 9(x - 3)^4 - 6(x - 3)^5$ be the fifth-degree Taylor polynomial for the function f about 3. Assume f has derivatives of all orders for all real numbers.

(A) Find $f(3)$ and the value of the fourth derivative, $f^{IV}(3)$.

(B) Write the third-degree Taylor polynomial for f' about 3, and use it to approximate $f'(3.2)$.

(C) Write the sixth-degree Taylor polynomial for $h(x) = \int_3^x f(t)\, dt$ about 3.

(D) Can $f(4)$ be determined from the given information? Explain.

3. Let R be the region enclosed by the graphs of $y = 4 - \frac{1}{2}x^2$ and $y = \sec\frac{x}{2}$.

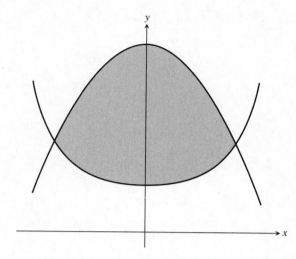

(A) Find the area of R.

(B) Find the volume of the solid generated when R is rotated about the x-axis.

(C) Write an expression involving one or more integrals that gives the length of the boundary of region R. DO NOT EVALUATE.

Calculus BC—Exam 1
Section II, Part B

Time: 45 minutes
Number of problems: 3

NO CALCULATOR MAY BE USED IN THIS PART OF THE EXAMINATION.

4.   Consider the curve given by $4x - 2x^2y = y^2 + 1$.

   (A)   Show that $\dfrac{dy}{dx} = \dfrac{2 - 2xy}{y + x^2}$.

   (B)   Show there is a point Q with x-coordinate 1 such that there is a horizontal tangent to the curve at Q.

   (C)   Evaluate $\dfrac{d^2y}{dx^2}$ at point Q. Is there a local minimum, local maximum, or neither at Q? Justify your answer.

5. The graph of g shown in the figure consists of a quarter-circle and three line segments. Let h be the function defined by

$$h(x) = \int_0^x g(t)\, dt.$$

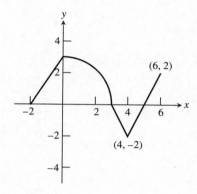

(A) Evaluate $h(4)$.

(B) Find all values of x in the interval $[-2, 6]$ at which h has a relative minimum. Justify your answer.

(C) Find the value of $h'(2)$.

(D) Find the x-coordinate of each point of inflection of the graph of h on the interval $(-2, 6)$. Justify your answer.

6. Consider the differential equation $\dfrac{dy}{dx} = x^2(1-y)^2$.

(A) On the axes provided, sketch a slope field for the given differential equation at the 15 points indicated.

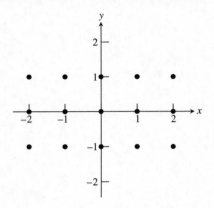

(B) Find the particular solution $y = f(x)$ to the given differential equation if $f(3) = 0$.

(C) For $f(x)$ found in part (B), find $\lim\limits_{x\to\infty} f(x)$. Explain how the answer is related to a characteristic of the slope field.

Calculus BC—Exam 2

Section I, Part A

Time: 55 minutes
Number of questions: 28

NO CALCULATOR MAY BE USED IN THIS PART OF THE EXAMINATION.

Directions: Solve each of the following problems. After examining the form of the choices, decide which is the best of the choices given.

In this test: Unless otherwise specified, the domain of a function f is assumed to be the set of all real numbers x for which $f(x)$ is a real number.

1. What are all values of x for which the function f defined by $f(x) = 2x^3 - 3x^2 - 36x - 4$ is decreasing?

    (A) $-2 < x < 3$ (B) $-3 < x < 2$

    (C) $x < -2$ or $x > 3$ (D) $x < -3$ or $x > 2$

    (E) All real numbers

2. In the xy-plane, the graph of the parametric equations $x = 3t - 1$ and $y = -2t + 4$, for $-5 \le t \le 5$ is a line segment with slope

    (A) -2.   (B) $-\dfrac{3}{2}$.   (C) $-\dfrac{2}{3}$.

    (D) 3.   (E) 6.

3. The slope of the line tangent to the curve $xy + (y + 1)^2 = 6$ at the point $(2, 1)$ is

    (A) -4.   (B) $-\dfrac{2}{5}$.   (C) $-\dfrac{1}{6}$.

    (D) $\dfrac{1}{8}$.   (E) 8.

4. $\int \dfrac{1}{x^2 + 3x - 10}\,dx =$

(A) $-\dfrac{1}{7}\ln|(x + 5)(x - 2)| + C$

(B) $\dfrac{1}{7}\ln|(x + 5)(x - 2)| + C$

(C) $\ln|(x + 5)(x - 2)| + C$

(D) $\dfrac{1}{7}\ln\left|\dfrac{x - 5}{x + 2}\right| + C$

(E) $\dfrac{1}{7}\ln\left|\dfrac{x - 2}{x + 5}\right| + C$

5. If f and g are differentiable functions and if $h(x) = f(g(x^2))$, then $h'(x) =$

(A) $f'(g(x^2))g'(x^2).$ (B) $f'(g(x^2))g'(x^2)2x.$

(C) $f'(x^2)g'(x^2).$ (D) $f'(g'(x^2)).$

(E) $f'(g'(2x)).$

6. The graph of $y = h(x)$ is shown below. Which of the following could be the graph of $y = h'(x)$?

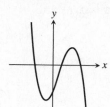

(A)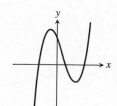

(B)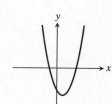

(C)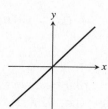

(D)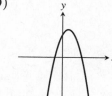

(E)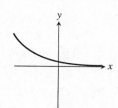

7. $\int_1^e \dfrac{3x^3 - 1}{x}\, dx =$

(A) $e^3 - 2$

(B) $e^3 - \dfrac{1}{e}$

(C) $\dfrac{9e^2 - 1}{e}$

(D) $\dfrac{3e^3 - 1}{e}$

(E) $\dfrac{2}{e}$

8. If $\dfrac{dy}{dx} = \cos x \sin^2 x$ and if $y = 1$ when $x = \dfrac{\pi}{2}$, what is the value of y when $x = 0$?

(A) $-\dfrac{2}{3}$

(B) $-\dfrac{1}{3}$

(C) 0

(D) $\dfrac{1}{3}$

(E) $\dfrac{2}{3}$

9. The rate of water pumped from a water storage tank is given by the graph shown. Of the following, which best approximates the total number of gallons of water pumped for the 24-hour period?

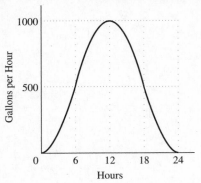

(A) 3000

(B) 6000

(C) 9000

(D) 12,000

(E) 15,000

10. A particle moves on a plane curve so that any time $t > 0$ its x-coordinate is $t^3 - 1$ and its y-coordinate is $(2t + 1)^3$. What is the acceleration vector of the particle when $t = 1$?

(A) $(0, 27)$

(B) $(6, 18)$

(C) $(6, 72)$

(D) $(6, 36)$

(E) $(10, 42)$

11. If the derivative of f is a constant and $0 < a < b$, then $\int_a^b f''(x)\, dx =$

(A) 0.

(B) 1.

(C) $\dfrac{ab}{2}$.

(D) $b - a$.

(E) $\dfrac{b^2 - a^2}{2}$.

12. If $f(x) = \begin{cases} e^x, & x \leq 2 \\ xe^2, & x > 2 \end{cases}$, then $\lim\limits_{x \to 2} f(x)$ is

(A) 1.   (B) e.   (C) e^2.

(D) $2e^2$.   (E) nonexistent.

13. The graph of the function f shown in the figure has a vertical tangent at the point $(1, 0)$ and a horizontal tangent at the point $(3, -2)$. For what values of x, $-2 < x < 5$, is f not differentiable?

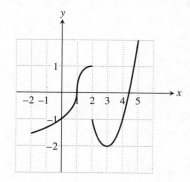

(A) 0 only   (B) 1 only   (C) 2 only

(D) 1 and 2 only   (E) 1, 2, and 3

14. What is the approximation of the value of cos 1 obtained by using the fourth-degree Taylor polynomial about $x = 0$ for cos x?

(A) $1 - \dfrac{1}{2} + \dfrac{1}{4}$ (B) $1 - \dfrac{1}{2} + \dfrac{1}{24}$ (C) $1 - \dfrac{1}{3} + \dfrac{1}{5}$

(D) $1 - \dfrac{1}{6} + \dfrac{1}{120}$ (E) $1 - \dfrac{1}{4} + \dfrac{1}{8}$

15. $\int x \sin x \, dx =$

(A) $-x \cos x + \sin x + C$ (B) $x \cos x + \sin x + C$

(C) $-x \sin x + \sin x + C$ (D) $x \sin x + \sin x + C$

(E) $-\dfrac{x^2}{2} \cos x + C$

16. If $f(x) = 3x^5 - 5x^4 + 2x - 1$, what are the x-coordinates of all points of inflection for the graph of f?

(A) -1 (B) 0 (C) 1

(D) 0 and 1 (E) $-1, 0$, and 1

17. The graph of a twice-differentiable function f is shown. Which of the following is true?

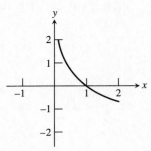

(A) $f(1) < f'(1) < f''(1)$ (B) $f'(1) < f(1) < f''(1)$

(C) $f''(1) < f(1) < f'(1)$ (D) $f''(1) < f'(1) < f(1)$

(E) $f''(1) < f(1) < f'(1)$

18. Which of the following series converge?

I. $\displaystyle\sum_{n=1}^{\infty} \frac{n}{n+3}$

II. $\displaystyle\sum_{n=1}^{\infty} \frac{\sin\left[(2n-1)\frac{\pi}{2}\right]}{n}$

III. $\displaystyle\sum_{n=1}^{\infty} \frac{2}{n}$

(A) None   (B) II only   (C) III only

(D) I and II only   (E) I and III only

19. The area of the region inside the polar curve $r = 4\cos\theta$ and outside the polar curve $r = 2$ is given by

(A) $\displaystyle\int_{-\pi/4}^{\pi/4}\left(4\cos\theta - 2\right)d\theta$

(B) $\displaystyle\int_{-\pi/3}^{\pi/3}\left(4\cos\theta - 2\right)^2 d\theta$

(C) $\displaystyle\frac{1}{2}\int_{-\pi/3}^{\pi/3}\left(4\cos\theta - 2\right)^2 d\theta$

(D) $\displaystyle\int_{0}^{\pi/3}\left(4\cos\theta - 2\right)^2 d\theta$

(E) $\displaystyle\int_{0}^{\pi/3}\left(16\cos^2\theta - 4\right)d\theta$

20. When $x = 16$, the rate at which $\sqrt{x}$ is increasing is $1/k$ times the rate at which x is increasing. What is the value of k?

(A) 2   (B) 4   (C) 6

(D) 8   (E) 16

21. The length of the path described by the parametric equations $x = t^2$ and $y = 3t + 1$, where $0 \le t \le 2$, is given by

(A) $\displaystyle\int_0^2 \sqrt{4t^2 + 9}\, dt.$

(B) $\displaystyle\int_0^2 \sqrt{t^2 + 3t + 1}\, dt.$

(C) $\displaystyle\int_0^2 \sqrt{2t + 3}\, dt.$

(D) $\displaystyle\int_0^2 t^2(3t + 1)\, dt.$

(E) $\displaystyle\int_0^2 (2t + 3)\, dt.$

22. If $\displaystyle\lim_{b\to\infty}\int_1^b \frac{1}{x\sqrt{x}}\, dx$ is finite, then which of the following must be true?

(A) $\displaystyle\sum_{n=1}^{\infty} n\sqrt{n}$ converges.

(B) $\displaystyle\sum_{n=1}^{\infty} \frac{1}{n\sqrt{n}}$ diverges.

(C) $\displaystyle\sum_{n=1}^{\infty} \frac{1}{n\sqrt{n}}$ converges.

(D) $\displaystyle\sum_{n=1}^{\infty} \frac{n}{\sqrt{n}}$ diverges.

(E) $\displaystyle\sum_{n=1}^{\infty} \frac{n}{\sqrt{n}}$ converges.

23. Let f be a function defined and continuous on the closed interval $[a, b]$. If f has a relative minimum at c and $a < c < b$, which of the following statements must be true?

    I. $f'(c)$ exists.

    II.   If $f'(c)$ exists, then $f'(c) = 0$.

    III. If $f''(c)$ exists then $f''(c) \geq 0$.

(A) II only   (B) III only   (C) I and II only

(D) I and III only   (E) II and III only

24. Which of the following differential equations generates the slope field shown in the figure?

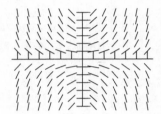

(A) $\dfrac{dy}{dx} = xy$ (B) $\dfrac{dy}{dx} = x$ (C) $\dfrac{dy}{dx} = y$

(D) $\dfrac{dy}{dx} = x + y$ (E) $\dfrac{dy}{dx} = x^2$

25. $\displaystyle\int_0^\infty \frac{3x^2}{(1 + x^3)^2}\, dx$ is

(A) -1.   (B) 0.   (C) 1.

(D) 3.   (E) divergent.

26. The population $P(t)$ of a species satisfies the logistic differential equation $\frac{dp}{dt} = p\left(4 - \frac{p}{1000}\right)$, where the initial population is 500 and t is the time in years. What is $\lim_{t\to\infty} p(t)$?

(A) 250   (B) 1000   (C) 1400

(D) 4000   (E) 10,000

27. If $\sum_{n=0}^{\infty} a_n x^n$ is a Taylor series that converges to $f(x)$ for all real x, then $f'(x) =$

(A) 0.   (B) a_1.   (C) $\sum_{n=0}^{\infty} na_n x^{n-1}$.

(D) $\sum_{n=0}^{\infty} na_n^{n-1}$.   (E) nx^{n-1}.

28. $\lim_{x\to 1} \dfrac{\int_1^x \sin t \, dt}{x^2 - 1}$ is

(A) 0.   (B) 1.   (C) $\frac{1}{2}$.

(D) 2.   (E) nonexistent.

Calculus BC—Exam 2
Section I, Part B

Time: 50 minutes
Number of questions: 17

A GRAPHING CALCULATOR IS REQUIRED FOR SOME QUESTIONS IN THIS PART OF THE EXAMINATION.

Directions: Solve each of the following problems. After examining the form of the choices, decide which is the best of the choices given.

In this test:

1. The exact numerical value of the correct answer does not always appear among the choices given. When this happens, select from among the choices the number that best approximates the exact numerical value.
2. Unless otherwise specified, the domain of a function f is assumed to be the set of all real numbers x for which $f(x)$ is a real number.

29. For what values of k will $\sum_{n=1}^{\infty} \left(\dfrac{k}{3}\right)^n$ converge?

    (A) $k > 3$ (B) $k < -3$ (C) $k < 3$

    (D) $-3 \leq k \leq 3$ (E) $-3 < k < 3$

30. If f is a vector-valued function defined by $f(t) = (e^{2t}, \sin t)$, then $f''(t) =$

    (A) $(2e^{2t}, \cos t)$.   (B) $(4e^{2t}, -\sin t)$.   (C) $(e^{2t}, \sin t)$.

    (D) $(4e^{2t}, \sin t)$.   (E) $(4e^{2t}, \cos t)$.

31. The radius of a circle is increasing at a constant rate of 0.2 centimeters per second. In terms of the circumference C, what is the rate of change of the area of the circle, in square centimeters per second?

    (A) 0.2 (B) $2\pi(0.2)$ (C) $0.2C$

    (D) $\pi(0.2)^2$ (E) $\pi(0.2)^2 C$

32. Let f be the function given by $f(x) = \dfrac{(x-1)(x^2-9)}{x^2-a}$. For what positive values of a is f continuous for all real numbers x?

(A) None   (B) 1 only   (C) 3 only

(D) 9 only   (E) 1 and 3 only

33. Let R be the region enclosed by the graph of $y = \sin e^{x^2}$, the x-axis, and the lines $x = -1$ and $x = 1$. What is the closest integer approximation of the area of R?

(A) 0   (B) 1   (C) 2

(D) 3   (E) 4

34. If $\dfrac{dy}{dx} = \sqrt{y^2 + 1}$, then $\dfrac{d^2y}{dx^2} =$

(A) $2y$.   (B) $-y$.   (C) $\dfrac{y}{\sqrt{y^2+1}}$.

(D) y.   (E) $\sqrt{2y}$.

35. If $f(x) = 2g(x) - 1$ for $1 \le x \le 3$, then $\int_1^3 (f(x) + g(x))\, dx =$

(A) $\int_1^3 g(x)\, dx - 2$.   (B) $2\int_1^3 g(x)\, dx - 2$.

(C) $3\int_1^3 g(x)\, dx - x$.   (D) $3\int_1^3 g(x)\, dx - 2$.

(E) $2\int_1^3 g(x)\, dx - 1$.

36. The Taylor series for e^x, centered at $x = 0$, is $\displaystyle\sum_{n=0}^{\infty} \frac{x^n}{n!}$. Let f be the function given by the sum of the first four nonzero terms of this series. The maximum value of $|e^x - f(x)|$ for $-0.6 \le x \le 0.6$ is

(A) 0.0036.   (B) 0.0048.   (C) 0.0052.

(D) 0.0061.   (E) 0.0081.

37. What are all values of x for which the series $\sum_{n=1}^{\infty} \dfrac{(x-1)^n}{3^n}$ converges?

   (A) $-2 < x < 4$ (B) $-2 \le x < 4$ (C) $-2 \le x < 4$

   (D) $-1 < x < 1$ (E) $-1 \le x \le 1$

38. The function f is continuous on the closed interval $[1, 7]$ and has values that are given in the table. Using the subintervals $[1, 4]$, $[4, 6]$, and $[6, 7]$, what is the trapezoidal approximation of $\int_1^7 f(x)\, dx$?

x	1	4	6	7
$f(x)$	20	40	50	30

   (A) 110   (B) 130   (C) 180

   (D) 210   (E) 220

39. The base of a solid is a region in the first quadrant bounded by the x-axis and the curve $y = \sin x$, where $0 \le x \le \pi$, as shown in the figure. If cross sections of the solid perpendicular to the x-axis are squares, what is the volume of the solid?

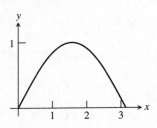

   (A) 0.785   (B) 1   (C) 1.571

   (D) 2   (E) 2.14

40. Which of the following is an equation of the line tangent to the graph of $f(x) = \dfrac{x^4}{4} - x^3$ at the point where $f'(x) = 1$?

   (A) $y = x + 3.5954$ (B) $y = x - 9.803$

   (C) $y = x - 3.056$ (D) $y = x - 1$

   (E) $y = 3x - 1$

41. Let $g(x) = \int_0^x f(t)\, dt$, where $0 \le x \le 6$. The figure below shows the graph of g on $0 \le x \le 6$. Which of the following could be the graph of f on $0 \le x \le 6$?

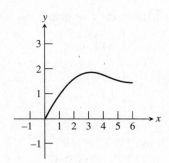

(A)

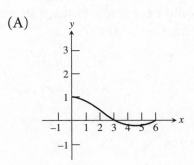

(B)

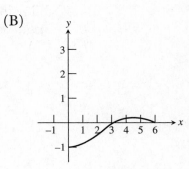

(C)

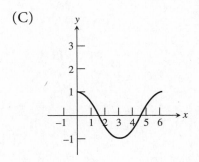

(D)

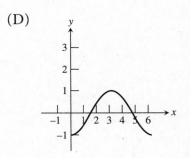

(E)

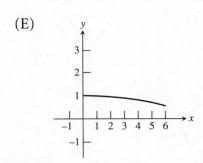

42. The graph of the function represented by the Maclaurin series

$$-x + \frac{x^3}{3!} - \frac{x^5}{5!} + \frac{x^7}{7!} - \cdots + (-1)^{n+1}\frac{x^{2n+1}}{(2n+1)!} + \cdots$$

intersects the graph of $y = x^3 - 2$ at $x =$

(A) 1.043.   (B) 1.312.   (C) 1.441.

(D) 1.581.   (E) 1.731.

43. A particle starts from rest at the point $(3, 0)$ and moves along the x-axis with a constant negative acceleration for time $t \geq 0$. Which of the following could be a graph of the distance $s(t)$ of the particle from the origin as a function of time?

(A)
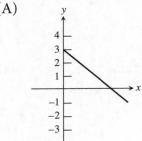

(B)
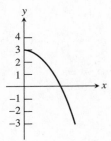

(C)
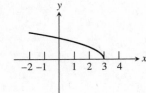

(D)
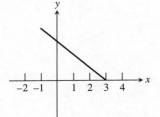

(E)
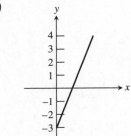

44. The data for the acceleration $a(t)$ of a car from 0 to 9 seconds are given in the table. If the velocity at $t = 0$ is 8 feet per second, the approximate value of the velocity at $t = 9$, computed using a left-hand Riemann sum with three subintervals of equal length, is

t (sec)	0	3	6	9
$a(t)$	4	1	7	3

(A) 20 ft/sec.   (B) 28 ft/sec.   (C) 36 ft/sec.

(D) 40 ft/sec.   (E) 44 ft/sec.

45. Let f be the function given by $f(x) = x^2 - 3x + 5$. The tangent line to the graph of f at $x = 3$ is used to approximate values of $f(x)$. Which of the following is the greatest value of x for which the absolute value of the error resulting from this tangent line approximation is less than 0.5?

(A) 3.4   (B) 3.5   (C) 3.6

(D) 3.7   (E) 3.8

Calculus BC—Exam 2
Section II, Part A

Time: 45 minutes
Number of problems: 3

A GRAPHING CALCULATOR IS REQUIRED FOR SOME PROBLEMS IN THIS PART OF THE EXAMINATION.

1. The power consumed in a house during a 24-hour period is described by

$$K(t) = 25 - 20 \cos \frac{\pi t}{12}$$

where $K(t)$ is measured in kilowatts and t is measured in hours.

(A) Sketch the graph of K on the grid provided.

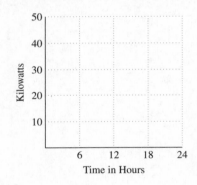

(B) Find the average number of kilowatts between $t = 6$ and $t = 18$.

(C) Peak usage is considered to be 35 kilowatts or higher. For what values of t is the number of kilowatts above 35?

(D) If power costs \$0.07 per kilowatt-hour, what is the total cost of power during the peak usage period?

2. Consider the differential equation given by $\dfrac{dy}{dx} = x^2 y$.

(A) On the axes provided, sketch a slope field for the given differential equation at the nine points indicated.

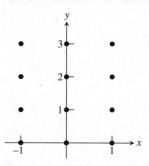

(B) Let $y = f(x)$ be the particular solution to the given differential equation with the initial condition $f(0) = 1$. Use Euler's method starting at $x = 0$, with a step size of 0.1, to approximate $f(0.2)$. Show the work that leads to your answer.

(C) Find the particular solution $y = f(x)$ to the given differential equation with the initial condition $f(0) = 1$. Use your solution to find $f(0.2)$.

3. Let f be a function that has derivatives of all orders for all real numbers. Assume $f(0) = 7, f'(0) = -4, f''(0) = 1$, and $f'''(0) = 6$.

(A) Write the third-degree Taylor polynomial for f about $x = 0$ and use it to approximate $f(0.3)$.

(B) Write the fourth-degree Taylor polynomial for g, where $g(x) = f(x^2)$, about $x = 0$.

(C) Write the fourth-degree Taylor polynomial for h, where $h(x) = \int_0^x f(t)\, dt$, about $x = 0$.

(D) Let h be defined as in part (C). Given that $f(1) = 5$, either find the exact value of $h(1)$ or explain why it cannot be determined.

Calculus BC—Exam 2
Section II, Part B

Time: 45 minutes
Number of problems: 3

NO CALCULATOR MAY BE USED IN THIS PART OF THE EXAMINATION.

4. Let R be the region in the first quadrant bounded by the graph of $y = 4 - x^2$, the x-axis, and the y-axis.

    (A)   Find the area of the region R.

    (B)   Find the volume of the solid generated when R is revolved about the x-axis.

    (C)   Find the volume of the solid generated when R is revolved about the y-axis.

5. Let f be the function defined by $f(x) = 2x \ln (2x)$.

    (A)   What is the domain of $f(x)$?

    (B)   Find the absolute minimum value of f. Justify that your answer is an absolute minimum.

    (C)   What is the range of f?

    (D)   Consider the family of functions defined by $y = bx \ln bx$, where b is a nonzero constant. Show that the absolute minimum value of $bx \ln bx$ is the same for all nonzero values of b.

6. A particle moves along the curve defined by the equation
    $y = x^2 - 2x$. The x-coordinate of the particle, $x(t)$, satisfies the equation $\dfrac{dx}{dt} = \dfrac{1}{\sqrt{t+1}}$, for $t \geq 3$ with the initial condition $x(3) = -1$.

    (A)   Find $x(t)$ in terms of t.

    (B)   Find dy/dt in terms of t.

    (C)   Find the location and speed of the particle at time $t = 8$.

Part V

Answers and Solutions

Part II: Precalculus Review of Calculus Prerequisites

Functions

Content & Practice

1. (A) $x \geq -4$

   (B) Range: {real numbers ≥ 0}. Since $x + 4 \geq 0$, $\sqrt{x + 4} \geq 0$, therefore $f(x) \geq 0$.

   (C) $(0, 2)$.   To find a y-intercept, let $x = 0$. $f(0) = 2$.

   (D) $(-4, 0)$.   To find x-intercepts, let $f(x) = 0$. $\sqrt{x + 4} = 0 \Rightarrow (x + 4) = 0 \Rightarrow x = -4$.

   (E)

   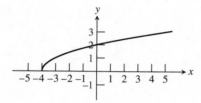

2. (A) All reals

   (B) Range: {real numbers ≥ -4.} The graph of $f(x)$ is a parabola with vertex $(1, -4)$. Since the parabola opens upward, the ordinate of the vertex is the least value of the function, -4.

   (C) $(0, -3)$. y-intercept: let $x = 0, f(0) = -3$

   (D) x-intercepts: $(3, 0)$ and $(-1, 0)$. Set $f(x) = 0$.

   $x^2 - 2x - 3 = (x - 3)(x + 1) = 0 \Rightarrow x - 3 = 0, x + 1 = 0 \Rightarrow x = 3, x = -1$

   (E)

   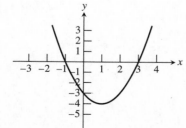

Additional Practice

1. (D) $\dfrac{1}{\sqrt{x-2}} \geq 0 \Rightarrow (x-2) > 0 \Rightarrow x > 2$

2. (i) $f(x) = 1/x$; Domain: $x \neq 0$. Range: {real numbers $\neq 0$}.

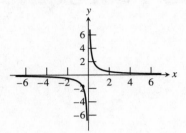

(ii) $g(x) = \sqrt{x}$; Domain: $x \geq 0$. Range: $h(x) \geq 0$.

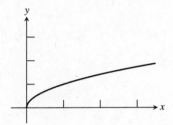

(iii) $h(x) = \dfrac{1}{x^2 - 9}$; Domain: $x \neq \pm 3$. Range: $h(x) \neq 0$.

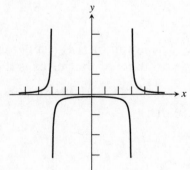

(iv) $k(x) = \dfrac{1}{\sqrt{x}}$; Domain: $x > 0$. Range: $k(x) > 0$.

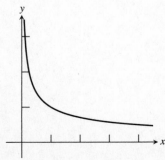

(v) $p(x) = x^2$; Domain: {reals}. Range: $P(x) \geq 0$.

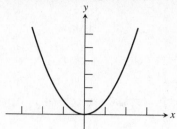

(vi) $q(x) = \sin x$; Domain: {reals}. Range: $[-1, 1]$.

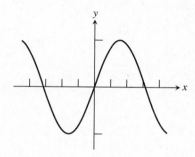

(vii) $s(x) = \tan x$; Domain: $\left\{ x \neq \dfrac{\pi}{2} + k\pi, ke \text{ integer} \right\}$. Range: {reals}.

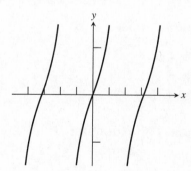

3. x-intercepts: $(0, 0)$ is both an x-intercept and a y-intercept, $(3, 0)$ and $(-3, 0)$ are additional x-intercepts.

$$f(x) = x^3 - 9x = x(x - 3)(x + 3) = 0 \Rightarrow x = 0, x = \pm 3$$

4.   (E). The range of a piecewise function is the union of the ranges of its separate parts. Since the range of $(x - 1)^2$, on its defined domain $x < 2$, is the set of nonnegative real numbers and the range of $2x - 3$ on $x > 2$, is reals >1, then the range of f must be $f(x) \geq 0$. Confirm with a graph.

Transformations

Content & Practice

1. (A) $x \geq -4$

   (B) Range: $g(x) \geq 3$. Since $x + 4 \geq 0$, $\sqrt{x + 4} \geq 0$, $2\sqrt{x + 4} + 3 \geq 3$. Therefore, $g(x) \geq 3$.

   (C) $g(x)$ is a transformation from $f(x)$: vertical stretch of 2, horizontal shift of -4, and vertical shift of 3.

   (D)

   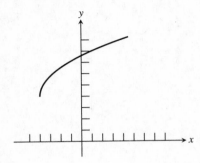

2. (A) $g(x) = -3(x - 1)^2 - 4$

   (B)

   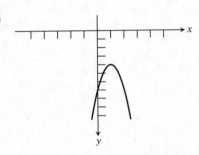

   (C) Range of function $g(x)$: all reals ≤ -4.

3. (B) $-f(x)$ produces an x-axis reflection of a function $f(x)$.

Additional Practice

1. (A) $y = -f(x - 1) - 2$, where $f(x) = |x|$

   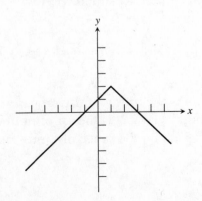

(B) $y = 2f(x/3 + 1)$, where $f(x) = |x|$

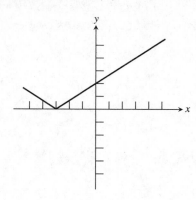

2. (D)

3. (B)

Polynomial Functions

Content & Practice

Selected answers are provided.

3. (A) 1 or 3 real zeros

(B) 0 or 2 extrema

(C) as $x \to \infty, y \to \infty$ and as $x \to -\infty, y \to -\infty$

(D) $\left\{ 1, \dfrac{7 \pm 3\sqrt{33}}{4} \right\}$.   The Rational Root Theorem helps us identify that 1 is root.

Dividing the given cubic by $(x - 1)$ produces the quadratic factor $(2x^2 - 7x - 31)$,

which yields the two additional roots.

(E) $(-1, 44)$ and $(4, -81)$

(F) Function $f(x)$ rising on both $(-\infty, -1]$ and $[4, \infty)$, and falling on $[-1, 4]$.

(G)

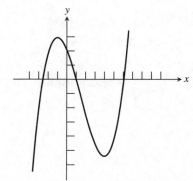

4. (B) $f(x)$ was expressed as the product of three distinct factors. Consider the zeros of each factor: $(3x + 1)$ has a single zero, $(x - 1)$ has a single zero (of multiplicity 3), and $x^2 + 4$ has no real zeros. Therefore, there are two distinct real zeros of $f(x)$.

Additional Practice

1. (A) 0, 2, or 4 real zeros

    (B) 1 or 3 extrema

    (C) as $x \to \pm \infty, f(x) \to \infty$

    (D) $\left\{ \pm 1, \pm \sqrt{7} \right\}$. $f(x)$ easily factors into $(x - 1)(x + 1)(x^2 - 7)$.

    (E) $(0, 7), (2, -9), (-2, -9)$

    (F) Function $f(x)$ rising on both $[-2, 0]$ and $[2, \infty)$, and falling on $(-\infty, -2]$ and $[0, 2]$.

    (G)

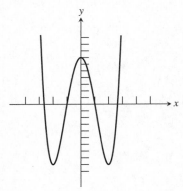

2. (E)

Rational Functions

Content & Practice

1. (A) $x \neq \pm 3$

    (B) $(3, 3/2)$. $f(x) = \dfrac{x^2 + 3x - 18}{x^2 - 9} = \dfrac{(x + 6)(x - 3)}{(x + 3)(x - 3)}$. The factor $(x - 3)$ appears in both the numerator and denominator exactly once.

    (C) $x = -3$. The factor $(x + 3)$ appears only in the denominator.

    (D) $y = 1$. The degree of the numerator equals the degree of the denominator, and the ratio of the two leading coefficients is 1. Therefore $y = 1$ will be a horizontal asymptote.

    (E) As $x \to \pm \infty, f(x) \to 1$.

    (F) $(0, 2)$

(G)

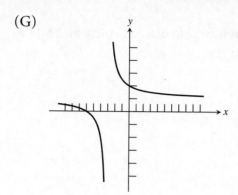

2. (A)

x	$f(x)$
0	0
0.9	-8.1
0.99	-98.01
0.999	-998
1.001	1002
1.01	102.01
1.1	12.1
2	4

(B) As $x \to 1^-$, the graph of f falls toward $-\infty$. As $x \to 1^+$, the graph of f rises toward $+\infty$.

Additional Practice

1. (A) $x \neq 1$

(B) No removable discontinuities. There are no common factors in the numerator and denominator to consider.

(C) $x = 1$

(D) $y = x + 2$ is a slant asymptote. The degree of the numerator is 1 greater than the degree of the denominator. The linear asymptote can be confirmed by graphing.

(E) $x \to \infty, y \to \infty$, and $x \to -\infty, y \to -\infty$

(F) $(0, -3)$

(G)

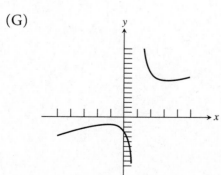

2. (E) $f(x) = \dfrac{x^2 - 2x}{x^2 - 4} = \dfrac{x(x - 2)}{(x + 2)(x - 2)}$ implies a removable discontinuity at 2 and a vertical asymptote at $x = -2$.

Exponential Functions

Content & Practice

1. (A) $(0, 28)$

    (B)

x	f(x)
−2	100
−1	52
1	16
2	10

    (C) f is decreasing. As x increases, the function value decreases because $0 < b < 1$ and $a > 0$.

    (D) $y = 4$

    (E) As $x \to +\infty$, $f(x) \to 4$. As $x \to -\infty$, $f(x) \to \infty$.

    (F)

    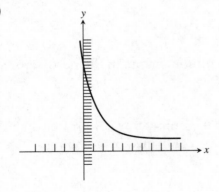

Additional Practice

1. (A) $(0, 3)$. Let $x = 0$, $f(0) = 3$.

    (B)

x	f(x)
−2	0.75
−1	1.5
1	6
2	12

(C) Increasing. As x increases, the values of $f(x)$ increases, because $b > 1$ and $a > 0$.

(D) $y = 0$

(E) As $x \to -\infty$, $f(x) \to 0$.

As $x \to +\infty$, $f(x) \to \infty$.

(F) (F)

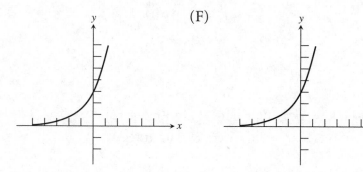

2. (B). $A(t) = 100\left(\frac{1}{2}\right)^{t/2}$. $A(4) = 25$ and $A(8) = 6.25$

Sinusoidal Functions

Content & Practice

1. (A) Amplitude is 3.

(B) Vertical shift of $+1$

(C) Range: $[-2, 4]$

(D) Horizontal shift of $\pi/4$ to the right.

(E) Period is 4π. $\dfrac{2\pi}{\frac{1}{2}} = 4\pi$.

(F) Local maxima determined by $\left(\dfrac{\pi}{4} + 4k\pi, 4\right)$, where k is an integer.

Local minima determined by $\left(\dfrac{9\pi}{4} + 4k\pi, -2\right)$, where k is an integer.

(G)

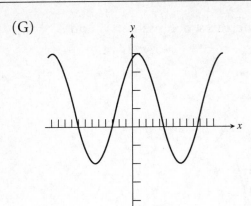

2.(A)   Local maxima determined by $(\pi + 2k\pi, 1)$, where K is an integer.
Local minima determined by $(2k\pi, -1)$, where K is an integer.

(B)   Amplitude is 1

(C)   Vertical shift: none

(D)   Period is 2π.

(E)   Example: $f(x) = -\cos x$

(F)   Example: $f(x) = \sin \left(x - \frac{\pi}{2} \right)$

Additional Practice

1.   (A) Amplitude is $\dfrac{(8 - {}^-2)}{2} = 5$.

(B)   Vertical shift: $+3$, $-2 + 5$, or $(8 + {}^-2)/2$

(C)   Range: $[-2, 8]$

(D)   Period $= (6 - 2) \cdot 2 = 8$.

(E)   Example: $y = 5 \cos \left[\dfrac{\pi}{4} (x - 2) \right] + 3$

(F)   Example: $y = 5 \sin \dfrac{\pi}{4} x + 3$

(G)

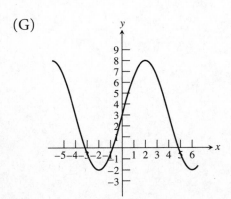

2. (A)

3. (E)

More Trigonometric Functions

Content & Practice

1. (A) Period: 2π

   (B) Domain: $x \neq \pi + 2k\pi$, where k is an integer.

   (C) Equations of vertical asymptotes: $x = (2k + 1)\pi$, where k is an integer.

   (D) $\ldots, (-2\pi, 0), (0, 0), (2\pi, 0), (4\pi, 0), \ldots$

   (E)

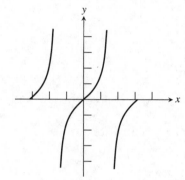

2. Vertical stretch by a factor of 2, vertical shift up 1; horizontal shift right π, horizontal shrink by 1/3

3. (A)

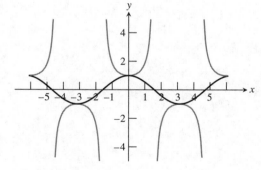

   (B)

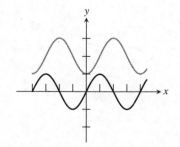

(C)

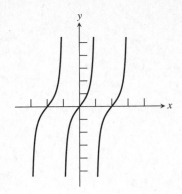

Additional Practice

1. (A) Period: π

    (B) Domain: $x \neq \dfrac{\pi}{4} + k \cdot \dfrac{\pi}{2}$, where k is an integer.

    (C) Vertical asymptotes: $x = \dfrac{\pi}{4} + 2k\pi$, where k is an integer.

    (D) Vertical shift: $+1$, where k is an integer.

    (E) Local minima determined by $\left(\dfrac{\pi}{2} + k\pi, -2 \right)$, where k is an integer.

    Local maxima determined by $(k\pi, 4)$, where k is an integer.

    (F)

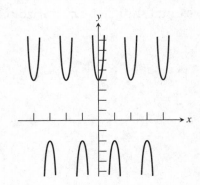

2. (D)

3. (D)

Inverse Trigonometric Functions

Content & Practice

1. (A) $\dfrac{5\pi}{6}$

    (B) $-\dfrac{\pi}{4}$

    (C) $\dfrac{\pi}{4}$

2. (A) 0.509

    (B) 1.176

    (C) -0.736

    (D) No solution

3. (A) $x = \pm\dfrac{\pi}{3} + 2k\pi$, where k is an integer.

    (B) $x = -\dfrac{\pi}{2} + 2k\pi$; where k is an integer.

    (C) $x = -\dfrac{\pi}{6} + k\pi$, where k is an integer.

Additional Practice

1. (A) 1.819
    (B) -0.983
    (C) 0.389

2. (A) $x = -\dfrac{\pi}{4} + k\pi$, where k is an integer. $3 + \tan x = 2 \Rightarrow \tan x = -1$

    (B) $x = \pm\dfrac{\pi}{6} + k\pi$; where k is an integer. $4\cos^2 x = 3 \Rightarrow \cos x = \pm\dfrac{\sqrt{3}}{2}$

    (C) $x = k\pi$ or $x = \dfrac{\pi}{6} + 2k\pi$ or $\dfrac{5\pi}{6} + 2k\pi$, where k is an integer.

    $2\sin^2 x = \sin x \Rightarrow \sin x(2\sin x - 1) = 0$

    (D) No solutions. $\cos^2 x = 4 \Rightarrow \cos x = \pm 2$ but $-1 \le \cos x \le 1$!

3. (A) $\sin^{-1} x$ is defined only on $\left[-\pi/2, \pi/2 \right]$, where it has the single solution, $-\pi/6$.

Parametric Relations

Content & Practice

1.

t	x	y
-2	6	-4
-1	2	-2
0	0	0
1	0	2
2	2	4
3	6	6

2. $x = t^2 - t, y = 2t; y = 2t \Rightarrow t = \frac{1}{2}y$. Therefore,

$$x = \left(\frac{1}{2}y\right)^2 - \left(\frac{1}{2}y\right)$$

$$x = \frac{1}{4}y^2 - \frac{1}{2}y$$

3. Utilizing the identity assuring that $\sin^2 t + \cos^2 t = 1$,

$$x = 3\sin t \Rightarrow \sin t = \frac{x}{3} \quad \text{and} \quad y = 4\cos t \Rightarrow \cos t = \frac{y}{4}.$$

$$\sin^2 t + \cos^2 t = \left(\frac{x}{3}\right)^2 + \left(\frac{y}{4}\right)^2 = 1 \quad \text{or} \quad \frac{x^2}{9} + \frac{y^2}{16} = 1, \text{ an ellipse}$$

4. $x_t = t \qquad\qquad t_{min} = 2$

$y_t = (t - 2)^2 \quad t_{max} = \infty$

$\qquad\qquad\qquad t_{step} = 0.1$

5. (A) A compared to B: The smaller parametric t_{step} in B caused the graph to be generated more slowly. This was because the calculator was generating about ten times as many points in graphing B.

(B) A compared to C: The negative t_{step} in C and the reversal of values in t_{min} and t_{max} caused the graph to be generated right to left, rather than left to right as it was in A. This was because the calculator begins generating points by using the t_{min} and incrementing that value by t_{step}.

6. Similarities: Plots D and E generate circles with centers $(0, 0)$. Differences: The plot D has radius 3 and is generated in a counterclockwise fashion beginning with the point $(3, 0)$, while plot E has radius 1 and is generated clockwise from $(0, 1)$. The radius differences were caused by the constant factor of 3 in D. The point generation differences were caused by the reversal of trig functions within x_t, y_t.

7. $x_t = 88 \cos \frac{\pi}{2}, \quad y_t = -16t^2 + 88t + 0$

$x_t = 0, \qquad\qquad y_t = -16t^2 + 88t$

8. $x_t = 15 \cos\left(-\frac{\pi}{10}(t+5)\right), y_t = 15 \sin\left(-\frac{\pi}{10}(t+5)\right) + 23$, with $t_{\min} = 0$ and $t_{\max} = 120$.

Radius is 15 feet, height off the ground is 8 feet, and the period is given as 20, so

$20 = \frac{2\pi}{|B|}, |B| = \frac{\pi}{10}$. For clockwise motion, $B < 0$, so $B = -\frac{\pi}{10}$.

Additional Practice

1. (B)   Eliminate the parameter: $x = 2t + 3 \Rightarrow t = \frac{x-3}{2} \Rightarrow y = \sqrt{\left(\frac{x-3}{2}\right) - 3}$, which is a

   portion (half) of the parabola $y^2 = \frac{1}{2}x - \frac{9}{2}$.

2. (A)   The parametric equations are $x = 48 \cos(35°)t$ and $y = -16t^2 + 48 \sin(35°)t + 5$.
   When the ball hits the ground, $y = 0$. Solving the second equation for t, setting $y = 0$, tells us
   that $t = 1.886$ seconds. Evaluating the first equation for x at 1.886 seconds indicates a distance
   of about 74 feet.

Numerical Derivatives and Integrals

Content & Practice

1. 14.778. $m_{\tan} = \frac{f(1.001) - f(0.999)}{1.001 - 0.999} = \frac{7.4038 - 7.3743}{0.002} = 14.778$

2. Exact Area $= 12$. Built-in graphing calculator functions give many approximate answers. The
   exact area is 12.

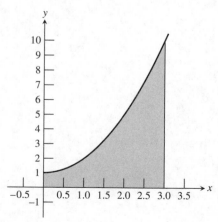

Additional Practice

1. **(D)** Numerical derivative gives a slope value of 4 and the point $(2, 6)$.
   $y - 6 = 4(x - 2) \Rightarrow y = 4x - 2$.

2. **(C)** Evaluating the numerical derivative of $h = 200 - 16t^2$ at $t = 2$ yields -64 feet/second.

3. **(A)** -20 gal/min.   Avg. $= \dfrac{G_4 - G_0}{t_4 - t_0} = \dfrac{40 - 120}{4 - 0} = -20$ gal/min.

   (B) -14 gal/min.   Avg. $= \dfrac{G_3 - G_{2.5}}{t_3 - t_{2.5}} = \dfrac{50 - 57}{3 - 2.5} = \dfrac{-7}{0.5} = -14$ gal/min.

Part III: Review of AP* Calculus AB and BC Topics Functions, Graphs, and Limits

Analysis of Graphs

Additional Practice

1. Domain: Since the argument $1 + 6x$ must be nonnegative, $x \geq -\frac{1}{6}, \left[-\frac{1}{6}, \infty\right)$.

   Range: Since $f\left(-\frac{1}{6}\right) = 0$ and $f \geq 0$, $[0, \infty)$

2. $\{-2, 2, 3\}$. Use the Rational Root Theorem, or Guess and Check, to find a single solution, e.g., any of the set $\{-2, 2, 3\}$. Once one solution has been found—say, 2—form a factor of $f(x), (x - 2)$. Polynomial long division will yield a quadratic factor that can be factored and solved to provide the remaining two solutions.

$$\text{or } f(x) = (x^3 - 3x^2)(4x - 12)$$
$$= x^2(x - 3) \cdot 4(x - 3)$$
$$= (x^2 - 4)(x - 3)$$
$$= (x - 2)(x + 2)(x - 3)$$

3. (A) Find the domain of S. $\{x: 0 \leq x < \pi/2 \text{ and } 3\pi/2 < x \leq 2\pi\}$

   $S(x) = g(f(x)) = \ln(\cos x)$. Remember that $f(x) = \cos x$ with $0 \leq x \leq 2\pi$ generates values between -1 and 1. Since the domain of $g(x) = \ln x$ can accept only positive values, restrict the domain of f to values for which $f(x) > 0$, namely, $0 \leq x < \pi/2$ and $3\pi/2 < x \leq 2\pi$.

   (B) Range: $\{y: -\infty < y \leq 0\}; (-\infty, 0]$

   For $0 \leq x < \pi/2$ and $3\pi/2 < x \leq 2\pi, 0 < \cos x \leq 1$. Therefore, $S(x) = \ln(\cos x)$ will generate only nonpositive values. (Think of the graph of $g(x) = \ln x$—it is below the x-axis for argument values between 0 and 1, and has value 0 at 1.) Therefore, $-\infty < S(x) \leq 0$.

   (C) Zeros: $\{0, 2\pi\}$

   $S(x) = 0$ whenever $\cos x = 1$. $\cos x = 1$ for $x = 0$ and $x = 2\pi$.

4. $\{-2, 2, 1\}$. Use the Rational Root Theorem, or Guess and Check, to find a single solution, e.g., any of the set $\{-2, 2, 1\}$. Once one solution has been found—say 2—form a factor of $f(x), (x - 2)$. Polynomial long division will yield a quadratic factor that can be factored and solved to provide the remaining two solutions.

5. (A) $y = |f(x)|$.

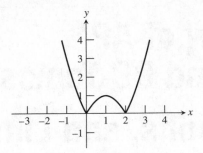

(B) $y = f(|x|)$.

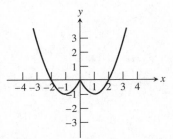

6. (A) Domain of H: $[0, \infty)$. Determination begins with consideration of domain and range of g, $x \geq 0$. The function f will accommodate the entire range of g, $g(x) \geq 1$. Therefore, the domain of H will match the domain of g.

$$H(x) = f(g(x)) = \ln(e^{2x})^2 = \ln e^{4x} = 4x, \quad \text{for } x \geq 0.$$

(B) Domain of K: $[1, \infty)$. Determination begins with consideration of domain of f, $x > 0$. The function g will only accommodate nonnegative values of f. f will be nonnegative for $x \geq 1$. Therefore, the domain of K is the subset of the domain of g, $x \geq 1$.

$$K(x) = g(f(x)) = e^{2(\ln x^2)} = e^{\ln x^4} = x^4, \quad \text{for } x > 1.$$

(C) $f^{-1}(x) = e^{x/2}$, $x \in \{\text{reals}\}$. $y = \ln(x^2)$, $x > 0$. Find the inverse by interchanging x and y and solving for y. $x = \ln y^2 \Rightarrow y = e^{x/2}$. Since the range of f includes all real numbers, so the domain of f^{-1}.

7. $\{\pm \pi/2, 0\}$. $0 = \cos x - \cos^2 x = \cos x(1 - \cos x) \Rightarrow x = \pm \pi/2, 0$.

8. (A) $\{\pm 1\}$. Note: $x = 0$ is not a zero because the function is not defined there.

(B) Vertical asymptote equations: $x = 2$ and $x = -2$. Horizontal asymptote: $y = 1$.

(C) Symmetric with respect to the y-axis because f is an even function.

9. f is an odd function. $f(-x) = \dfrac{-x + \sin -x}{\cos -x} = \dfrac{-x - \sin x}{\cos x} = -f(x)$.

10. (A) The graph of $f(x)$ is symmetric with respect to the y-axis, $f(-x) = f(x)$.

(B) Vertical asymptote equations: $x = 3$ and $x = -3$. Horizontal asymptote: $y = 9$, since $\lim\limits_{x \to \pm\infty} f(x) = 9$.

11. (B)

12. (C)

13. (E)

14. (A)

15. (E)

Limits of Functions

Content & Practice

1. $\lim\limits_{x \to 2} (x^3 - 5) = 3$

2. $\lim\limits_{x \to -4} \left(\dfrac{x^2}{x-2} \right) = -\dfrac{8}{3}$

3. (A) $\lim\limits_{x \to 3} \left(\dfrac{x^2 - 9}{x^2 - 5x + 6} \right) = 6$

(B) Substitution gives an answer of $\dfrac{0}{0}$.

(C) $\lim\limits_{x \to 3} \dfrac{x^2 - 9}{x^2 - 5x + 6} = \lim\limits_{x \to 3} \dfrac{(x-3)(x+3)}{(x-3)(x-2)} = \lim\limits_{x \to 3} \dfrac{x+3}{x-2} = 6$

(D)

x	$f(x)$
2.7	8.1429
2.8	7.25
2.9	6.5556
3	**6**
3.1	5.5455
3.2	5.1667

4. (A) $\lim\limits_{x \to -3^+} f(x) = -4$

(B) $\lim\limits_{x \to 8^-} f(x) = -3$

5. (A) $\lim\limits_{x \to -1^-} f(x) = -\infty$

   (B) $\lim\limits_{x \to -1^+} f(x) = \infty$

   (C) $\lim\limits_{x \to 2^-} f(x) = \infty$

   (D) $\lim\limits_{x \to 2^+} f(x) = -\infty$

   (E) $\lim\limits_{x \to -\infty} f(x) = 0$

   (F) $\lim\limits_{x \to +\infty} f(x) = 0$

Additional Practice

1. (A) $\lim\limits_{x \to 2^-} f(x) = 1$

   (B) $\lim\limits_{x \to 2^+} f(x) = 3$

   (C) $\lim\limits_{x \to 2} f(x)$ does not exist since the left-hand limit and the right-hand limit are not equal.

2. $\lim\limits_{x \to 2^-} f(x) = 1$ and $\lim\limits_{x \to 2^+} f(x) = 4 + a$. Therefore, $a = -3$.

3. (A) $\lim\limits_{x \to -\infty} f(x) = 0$

   (B) $\lim\limits_{x \to +\infty} f(x) = 0$

   (C) Conclusion: $y = 0$ is the horizontal asymptote.

4. (A) $\lim\limits_{x \to -\infty} f(x) = 2$

   (B) $\lim\limits_{x \to +\infty} f(x) = 2$

   (C) $\lim\limits_{x \to -2^-} f(x) = -\infty$

   (D) $\lim\limits_{x \to 2} f(x) =$ does not exist.

   (E) Conclusions: $y = 2$ is the horizontal asymptote, and $x = -2$ and $x = 2$ are vertical asymptotes.

5. (A) $\lim\limits_{x \to 0} f(x) = 2$

   (B) $\lim\limits_{x \to +\infty} f(x) = 0$

6. $\lim\limits_{x \to -1.8} f(x) \approx -22.60$

7. (D)

8. (D)

9. (C)

10. (D)

Asymptotic and Unbounded Behavior

Content & Practice

1. Vertical: $x = 4$; Horizontal: $y = 1$

2. Vertical: $x = 3$; Horizontal: none; Slant: $y = 2x + 9$

3. Vertical: none; Horizontal: $y = 0$

4. $h(x)$ in Problem 3

5. (B)

Additional Practice

1. (D)

2. (C)

Function Magnitudes and Their Rates of Change

Additional Practice

1. (C)

2. (C)

Continuity

Content & Practice

1. Discontinuous at $x = \dfrac{\pi}{2} + k\pi$, where k is an integer.

2. Discontinuous at x = -1.

3. Discontinuous at x = -1.

4. f is undefined wherever $\cos x = 0$.

5. g is undefined when its denominator is zero.

6. h is discontinuous at $x = -1$ because $\lim\limits_{x \to -1^-} h(x) = -5$ but $\lim\limits_{x \to -1^+} h(x) = -4$.

7. For f, part I fails and part II fails.

8. For g, part I fails.

9. For h, part 2 fails.

Additional Practice

1. Not continuous. $\lim\limits_{x \to 3} f(x)$ does not exist.

2. Not continuous. $\lim\limits_{x \to 0} g(x) = 5$, but $g(0) = 4$.

3. Continuous. $\lim\limits_{x \to 2} h(x) = 4$, and $h(2) = 4$.

4. (A) No. $\lim\limits_{x \to 3^-} f(x) = 1$, but $f(3) = 33$.

    (B) $b = \dfrac{1 - 9a}{3}$. $9a + 3b = 1 \Leftrightarrow b = \dfrac{1 - 9a}{3}$.

5. (E)

6. (A)

Intermediate and Extreme Value Theorems

Content & Practice

Additional Practice

1. We see that $f(x)$ is a continuous function on $[2, 3]$, so we can apply the IVT. Because $f(2)$ is positive and $f(3)$ is negative, the IVT tells us that between $x = 2$ and $x = 3$ the function has a value of zero.

2. Maximum 2.3, Minimum 0.5.

3. (A)

4. (B) and (D)

5. (C)

Parametric, Polar, and Vector Functions

Content & Practice

1. (A) $(0, 5)$

(B) The particle moves counterclockwise from initial point $(4, 0)$.

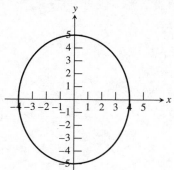

(C) 3 times

2. $\theta = \dfrac{\pi}{4} + k\pi \cdot r = 2\sin\theta, r = 2\cos\theta \Rightarrow \sin\theta = \cos\theta = \pm\dfrac{\sqrt{2}}{2}$, so $r = \pm\sqrt{2}$.

Thus $\left\{ (\theta, r): \theta = \dfrac{\pi}{4} + 2k\pi \text{ and } r = \sqrt{2}, \text{ or } \theta = \dfrac{5\pi}{4} + 2k\pi \text{ and } \right.$

$\left. r = -\sqrt{2} \text{ where } k \text{ is an integer.} \right\}$

3. (A) $\mathbf{r}(1) = \;<-1, 0>$ or $\mathbf{r}(1) = -\boldsymbol{i} + 0\boldsymbol{j}$

(B)

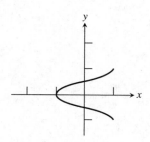

Additional Practice

1. (A)

2. (A)

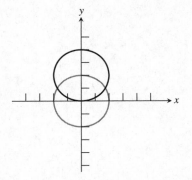

   (B) $(2, \pi/6), (2, 5\pi/6)$

3. Circle with radius 1. $x^2 + y^2 = \sin^2 2t + \cos^2 2t = 1.$

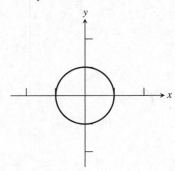

Derivatives

Concept of the Derivative

Content & Practice

1. (A) Linear function $\Rightarrow m = \dfrac{f(0) - f(-3)}{0 - (-3)} = -1$

    (B) Linear function $\Rightarrow m = \dfrac{f(3) - f(0)}{3 - 0} = 1$

2. $f'(550) = $ the rate at which profit is increasing (in dollars per basketball) when the number of basketballs being produced is 550.

Additional Practice

1. (E) A derivative is an *instantaneous* rate of change.

2. (D) A derivative is an *instantaneous* rate of change.

Concept of the Derivative

Content & Practice

1. (A) Differentiable at all domain values *except* $\{-2, 0, 2\}$. Continuous functions will not be differentiable at corners or cusps (or places where the tangent line is vertical).

    (B) At $x = -2$ and $x = 0$. Continuous functions will not be differentiable at corners or cusps.

    (C) At $x = 2$. The function is not continuous at that point, so it also fails to be differentiable at that point.

Additional Practice

1. (A) Since $f'(x)$ has a derivative at $x = 5$, it must also be continuous there.

2. (A) $\{-2, 2\}$. Jump discontinuity and removable discontinuity

    (B) $\{-3, -2, 2\}$. Discontinuous $\rightarrow$ nondifferentiable. The corner at $x = -3$ will cause the function to be not differentiable there either.

3. (C) A continuous function can still fail to be differentiable.

4.   (C)   A continuous function can still fail to be differentiable.

5.   (E)   Differentiability requires continuity, so a function that is not continuous will not be differentiable.

Slope of a Curve at a Point

Content & Practice

1.   (A) $m \approx 0$.   The tangent line at P is approximately horizontal.

(B) $m \approx 2$.   If the origin and point P are both on the tangent then $m = \dfrac{2 - 0}{1 - 0} = 2$.

(C)   Undefined.   The tangent line at P is approximately vertical.

2.   (D)   No tangent line exists at a corner.

3.   The left- and right-hand limits differ $\Rightarrow$ no two-sided limit $\Rightarrow$ no slope.

Additional Practice

1.   (A)   6.   Graph in window $[2, 5]$ by $[7, 20]$. The tangent line through $(3, 12)$ appears to pass through $(4, 8) \Rightarrow m = \dfrac{18 - 12}{4 - 3} = 6$.

(B) $m = \displaystyle\lim_{h \to 0} \dfrac{f(3 + h) - f(3)}{h} = \lim_{h \to 0} \dfrac{[(3 + h)^2 + 3] - (3^2 + 3)}{h} = \lim_{h \to 0}(6 + h) = 6$

2. $m = \dfrac{f(x + h) - f(x)}{h} = \dfrac{\dfrac{1}{x + h} - \dfrac{1}{x}}{h} = \dfrac{\dfrac{x}{x(x + h)} - \dfrac{x + h}{x(x + h)}}{h}$

$= \dfrac{\dfrac{-h}{x(x + h)}}{h} = \dfrac{-1}{x(x + h)} = -\dfrac{1}{x^2} \Rightarrow -\dfrac{1}{x^2} = -\dfrac{1}{4} \Leftrightarrow x = \pm 2$

3.   (D)   Because tangent lines at those points would be horizontal.

4.   (E)   Because the tangent line there is vertical.

Local Linearity

Content & Practice

1.   From the given information we know the slope is 0.5 and it passes through the point $(5, 3) \Rightarrow y - 3 = 0.5(x - 5) \Rightarrow y = 0.5x + 0.5$.

2. $f(5.023) \approx 0.5(5.023) + 0.5 = 3.0115$

Additional Practice

1.   From a calculator,
$f(2) = f'(2) \approx 7.389 \Rightarrow y - 7.389 = 7.389(x - 2) \Rightarrow y = 7.389x - 7.389.$

2.   (A) (i) Yes (flattens out as you zoom in)

   (ii) $f(1) = 2$ and $f'(1) = 2 \Rightarrow y - 2 = 2(x - 1) \Rightarrow y = 2x$

   (B) (i) Yes (slope $= 2$ everywhere, so differentiable)

   (ii) $y = 2x$. A linear function is its own tangent.

   (C) (i) Yes (flattens out if you zoom in)

   (ii) $f(1) = 2$ and $f'(1) = 2 \Rightarrow y - 2 = 2(x - 1) \Rightarrow y = 2x$

3.   (E) $f(0) = 3$ and $f'(0) = k \Rightarrow y - 3 = k(x - 0) \Rightarrow y = kx + 3.$ So,
$f(0.03) \approx k(0.03) + 3.$

4.   (A) $f(1) = a \ln 3 \Rightarrow$ point: $(1, a \ln 3); f'(1) = \dfrac{a}{3} =$ slope.

Tangent line:

$$y - a \ln 3 = \frac{a}{3}(x - 1) \Rightarrow y = a \ln 3 + \frac{a}{3}(x - 1) \Rightarrow f(0.98) \approx a \ln 3 + \frac{a}{3}(-0.02)$$

Instantaneous Rate of Change

Content & Practice

1.   (A) $\dfrac{f(1 + 5) - f(1)}{5} = \dfrac{108 - 3}{5} = 21$

   (B) $\dfrac{f(1 + 3) - f(1)}{3} = \dfrac{48 - 3}{3} = 15$

   (C) $\dfrac{f(1 + 1) - f(1)}{1} = \dfrac{12 - 3}{1} = 9$

   Approximations of $f'(1)$ will vary, but should reflect the trend in the answers above, so should be ≤ 9. Students who explore using smaller intervals should see a limit of 6 (the actual rate).

2. $\displaystyle\lim_{h \to 0} \frac{f(1 + h) - f(1)}{h} = \lim_{h \to 0} \frac{3(1 + h)^2 - 3(1)^2}{h} = \lim_{h \to 0}(6 + h) = 6.$ This is consistent with
our result in Problem 1.

3.   (B)   Slope at $x = 3$ is $\approx \dfrac{f(4) - f(2)}{4 - 2} = \dfrac{8.32 - 6.15}{4 - 2} = 1.085$.

Additional Practice

1. $\displaystyle\lim_{h \to 0} \dfrac{f(-2 + h) - f(-2)}{h} = \lim_{h \to 0} \dfrac{[6 - (-2 + h)^2] - [6 - (-2)^2]}{h} = \lim_{h \to 0}(4 - h) = 4$

2.   (A) $\dfrac{54.5 - 28.9}{2004 - 1994} = 2.56$

      (B) $\dfrac{44.8 - 33.3}{2002 - 1996} = 1.933$

      (C) $\dfrac{39 - 35.5}{2000 - 1998} = 1.75$

      (D)   1.75.   The best approximation is an average slope over the smallest interval we can get that includes the target value; that is, the slope from part (C).

3.   (E)   Sketch a tangent line at $x = 2$. It appears to pass through the points $(1, -4)$ and $(3, 4) \Rightarrow m = \dfrac{4 - (-4)}{2} = 4$.

Relationships between the Graphs of *f* and *f′*

Content & Practice

1.   (A) The values of f' are positive there.

      (B) The values of f' are negative there.

2.   Positive slopes indicate an increasing function, whereas negative slopes indicate a decreasing function.

3.   Whenever f' crosses the x-axis, f has an extreme point (maximum or minimum). f' represents the slope of f, so when it crosses the x-axis it means the slope of f is changing sign. Hence, f is changing from increasing to decreasing or vice versa, which creates an extreme value.

Additional Practice

1.   (D) f' has x-intercepts when the slope of f is zero (when f has a horizontal tangent line).

2.   (C) f' will be negative over intervals when f is decreasing.

3.   (D)   A, B, and C can all be ruled out simply because the slope of those graphs at their x-intercepts is clearly not zero (so the values of $f(x)$ and $f'(x)$ can't be equal). $f(x) = e^{x-5}$, on the other hand, is always positive and always increasing; the values of the function appear to match up with the slope.

The Mean Value Theorem

Content & Practice

1. (A) Yes; $f(x)$ is continuous on $[1, 5]$ and differentiable on $(1, 5)$.

    (B) $f'(x) = \lim\limits_{h \to 0} \dfrac{(x + h)^2 - x^2}{h} = \lim\limits_{h \to 0} \dfrac{2xh + h^2}{h} = \lim\limits_{h \to 0} (2x + h) = 2x$

    (C) $\dfrac{f(5) - f(1)}{5 - 1} = \dfrac{25 - 1}{4} = 6$

    (D) $2x = 6 \Rightarrow x = 3$

Additional Practice

1. (C) $f(x)$ is continuous on $[2, 7]$ and differentiable on $(2, 7)$.

2. $f(x)$ is continuous on $[2, 11]$ and differentiable on $(2, 11)$, so the conditions of the MVT are met. Graph $y = f'(x)$ and $y = \dfrac{\sin 11 - \sin 2}{11 - 2} \approx -0.212$ on the same axes and find the points of intersections over the interval $[2, 11]$. $\{4.499, 8.068, 10.782\}$

3. (D) The average rate of change over the interval $[-4, 4]$ is 0. For c to satisfy the MVT, $f'(c)$ must equal zero (that is, there must be a horizontal tangent at that point). This appears to happen around $x = 0$ and around $x = \pm 2$.

Equations Involving Derivatives

Content & Practice

1. $W'(t) = -3$ inches/hour. $W'(t)$ measures how the water level changes with respect to time; it is negative because the water is falling.

2. (C) A 10% increase indicates that the *rate of change* at any moment in time is equal to 10% of the current population (R).

3. The plane is descending at 200 feet per minute. The derivative measures the *rate of change* of the altitude, and a negative sign indicates the altitude is decreasing.

Additional Practice

1. $\dfrac{dV}{dt} = 8 \dfrac{dr}{dt}$ at $t = 5$. (Alternatively, $V'(5) = 8r'(5)$, with the presumption that the derivatives are taken with respect to time.)

2. $C'(w) = 23$. Here $C'(w)$ measures the *rate of change* of the cost, and the fact that it is positive shows the cost is increasing.

3. (C)   The statement involves the rate of change of the populations, which is $A'(t)$. "Directly proportional" indicates that it is some constant multiple (k) of the expression.

Correspondences Among the Graphs of *f*, *f'* and *f"*

Content & Practice

1. (A) Increasing.   (Slope is positive.)

    (B) Decreasing.   (Slope is negative.)

    (C) Local minimum.   (*f* is changing from decreasing to increasing.)

    (D) Local maximum.   (*f* is changing from increasing to decreasing.)

Additional Practice

1. (A)

    (B)

    (C) For the f' schematic, the sign is positive where the function is increasing and negative where it is decreasing. The zeros occur where the graph would have a horizontal tangent line. For the f'' schematic, the sign is positive where the graph is concave up, negative where it is concave down, and zero (or undefined) where there is a change of concavity (points of inflection).

2. (A) Approximately $[-3.1, 3.1]$.   In general, a function is increasing when it has a positive derivative. (Note: Even though the derivative is zero at the endpoints and at a single point within this interval, the graph is said to be increasing on the entire closed interval.)

    (B) Approximately $(-\infty, -3.1] \cup [3.1, \infty)$.   In general, a function is decreasing when it has a negative derivative. (Note: For a continuous function, the points where a function switches between increasing and decreasing are included in both intervals.)

    (C) At approximately $x = \pm 3.1$, where there is a change in sign in the derivative function. (There is a local *minimum* around $x = -3.1$ because the derivative (slope) is changing from negative to positive. There is a local *maximum* around $x = 3.1$ because the derivative (slope) is changing from positive to negative.)

3. (A) (a) Not only is there a change in sign in the second derivative, but since $f''(3)$ is defined, we know the first derivative exists and so there must be a tangent line at that point.

    (B) (b) There is a change in sign of the second derivative, but since $f''(5)$ is undefined, we can't be sure that a tangent line exists at that point.

    (C) (c) There is no change in sign in the second derivative.

4. Shown in bold is one possible graph of f. Variations include any vertical translation of the graph. Based on the information from f', we know f will be increasing until about $x = -0.3$, where it has a maximum, then decreasing until about $x = 2.4$, where it has a minimum, and then increasing again. From f'' we know that the graph of f will be concave down until about $x = 1$, and then it will be concave up.

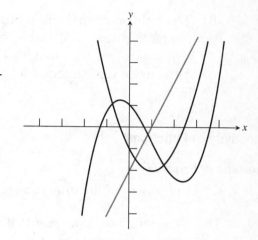

Points of Inflection

Content & Practice

1. (A) $f'(x) = -4x^3 + 12x^2 + 3 \Rightarrow f''(x) = -12x^2 + 24x$

    (B) $f''(x) = 0 \Rightarrow -12x^2 + 24x = 0 \Rightarrow -12x(x - 2) = 0 \Rightarrow x \in \{0, 2\}$. ($f''(x)$ is a polynomial function, so it is never undefined.)

    (C) $f''(x) = -12x(x - 2)$

$$f'' \xleftarrow[\quad\quad\underset{0}{\quad}\ \underset{2}{\quad}\quad\quad]{- - - -\ \ +\ +\ - -} $$

$f''(x)$ is positive on $(0, 2)$ and negative on $(-\infty, 0) \cup (2, \infty)$.

    (D) $(0, 5)$ and $(2, 27)$ f has points of inflection at $x = 0$ and $x = 2$ (f'' changes from negative to positive at $x = 0$, meaning that the graph changes from concave down to concave up; at $x = 2$, the concavity switches back again. We know f' exists at both of these points, so there is a tangent to the curve as well.)

Additional Practice

1. (B)

$$f'' \xleftarrow[\quad\quad\underset{-1}{\quad}\ \underset{2}{\quad}\quad\quad]{- - - -\ \ +\ +\ +\ +} $$

There is a point of inflection at $x = -1$, where the concavity changes from concave down to concave up.

2. $\left(\frac{1}{2}, -7\frac{1}{2}\right) f'(x) = 6x^2 - 6x + 6 \Rightarrow f''(x) = 12x - 6$. There is only one zero of $f''(x)$, at $x = \frac{1}{2}$. When $x < \frac{1}{2}, f''(x) < 0$ (making f concave down); when $x > \frac{1}{2}, f''(x) > 0$ (making f concave up). There is a point of inflection at $x = \frac{1}{2}$.

3. (A) (b)

   (B) The only point that could possibly be a point of inflection based on the information in the table is at $x = 5$, as it is the only place that $f''(x)$ changes signs (changing from negative to positive, thus making the graph change from concave down to concave up). Since $f'(5)$ has a value, there is a tangent line, and thus a point of inflection at that location.

Concavity of Functions

Content & Practice

1. Concave down on the intervals (A, C) and (E, G)

2. The x-coordinates of the points B, D, and F on the graph of $h'(x)$ would identify points of inflection on the graph of $h(x)$.

3. Yes. If f is any constant or linear function, $f''(x) = 0$ for all x yet there are no points of inflection.

4. Yes. The function $f(x) = x^{1/3}$ at $x = 0$ has a vertical tangent to the curve and $f''(0)$ is undefined.

5. Concave down on $(-\infty, -2) \cup (2, \infty)$

   $g(x) = \ln(4 + x^2) \Rightarrow g'(x) = \frac{2x}{4 + x^2} \Rightarrow g''(x) = \frac{2(4 - x^2)}{(4 + x^2)^2}. \quad g''(x) = 0 \Leftrightarrow x = \pm 2.$

   Analyzing the corresponding intervals leads to the answer.

6. (A)

Additional Practice

1. (D)

2. $h(x) = x^2 \cdot e^x \Rightarrow h'(x) = xe^x(x + 2) \Rightarrow h''(x) = e^x(x^2 + 4x + 2)$. Since $h'(-2) = 0$ and $h''(-2) < 0$, there is a relative maximum at $x = -2$.

Extreme Values of Functions

Content & Practice

1. A: local maximum; B: local and absolute minimum; C: local maximum; D: local minimum; E: local maximum and absolute maximum

2. Yes. It could be increasing (or decreasing) on both sides, e.g., $f(x) = x^3$ at $x = 0$.

3. Yes. f could be a function defined at $x = a$ with a vertical tangent, e.g., $f(x) = x^{2/3}$ at $x = 0$.

4. $5\frac{1}{3}$. Since $g'(x) = x^2 - 4$, the graph of g is decreasing on $[-1, 2]$ and increasing on $[2, 4]$. Therefore, either the left or right endpoint must be the absolute maximum. Evaluating $g(-1) = 3\,{}^2\!/_3$ and $g(4) = 5\,{}^1\!/_3$ shows that $x = 4$ generates the absolute maximum.

5. (B)

Additional Practice

1. (A)

2. (A) $f'\left(-\dfrac{1}{2}\right) = 0$ and f is increasing on the left and decreasing to the right (as far as $x = 1$).

   (B) Although $f'(-3) = 0$, f is increasing on both sides of $x = -3$.

   (C) f is increasing on the interval $[1, 5]$.

Analysis of Parametric, Polar, and Vector Curves

Content & Practice

1. (A) $\boldsymbol{v}(t) = \text{<}2t, 2\cos(2t)\text{>}$
   (B) $\boldsymbol{a}(t) = \text{<}2, -4\sin(2t)\text{>}$

   (C) $\dfrac{3}{\pi}$. $\dfrac{dy}{dx} = \dfrac{dy/dt}{dx/dt} = \dfrac{\cos 2t}{t}$. At $t = \dfrac{\pi}{6}$, $\dfrac{dy}{dx} = \dfrac{3}{\pi}$.

2. (A) $\boldsymbol{v}(t) = \text{<}(3t^2 - 1), 6(2t - 1)^2\text{>}$; $\boldsymbol{v}(1) = \text{<}2, 6\text{>}$

   (B) $\boldsymbol{a}(t) = \text{<}6t, 24(2t - 1)\text{>}$. When $(2t - 1)^3 = 0$, $t = \dfrac{1}{2}$. $\boldsymbol{a}\left(\dfrac{1}{2}\right) = \text{<}3, 0\text{>}$.

3. $\dfrac{dy}{dx} = \dfrac{-\sin(\theta)\sin(\theta) + [1 + \cos(\theta)]\cos(\theta)}{-\sin(\theta)\cos(\theta) - \sin(\theta)[1 + \cos(\theta)]} = 0$

   $-\sin(\theta)\sin(\theta) + [1 + \cos(\theta)]\cos(\theta) = 0$

   $-\sin^2(\theta) + \cos(\theta) + \cos^2(\theta) = 0$

$$\cos^2(\theta) - 1 + \cos(\theta) + \cos^2(\theta) = 0$$

$$2\cos^2(\theta) + \cos(\theta) + 1 = 0$$

$$[2\cos(\theta) - 1][\cos(\theta)] = 0$$

$$\cos(\theta) = \frac{1}{2} \text{ or } \cos(\theta) = -1$$

$$\theta = \left\{ \frac{\pi}{3}, \pi, \frac{5\pi}{3} \right\}.$$

We reject $\theta = \pi$ since it produces a zero in the denominator of $\frac{dy}{dx}$, so the final answer is $\theta = \left\{ \frac{\pi}{3}, \frac{5\pi}{3} \right\}.$

Additional Practice

1. (A) $\left\langle 0, -\frac{1}{t^2} \right\rangle$

   (B) Speed is $2\sqrt{2}$. Speed is $|\mathbf{v}(t)|$. Since $\mathbf{v}(t) = \left\langle 2, \frac{1}{t} \right\rangle$, the speed at $t = \frac{1}{2}$ is $2\sqrt{2}$.

2. (B) $\frac{dy}{dx}$, a rational function, is infinite when $\frac{dx}{dt} = 0$ but $\frac{dy}{dt} \neq 0$. $\frac{dx}{dt} = 0$ for $t = -3$ and $t = 2$.

3. (D) Particle moves left when $\frac{dx}{dt} < 0$.

Optimization

Content & Practice

1. Dimensions $2\sqrt[3]{3}$ by $2\sqrt[3]{3}$ by $\sqrt[3]{3}$ will minimize the surface area. Let the box dimensions be x by x by h. Volume $V = x^2h$ and surface area $S = x^2 + 4xh$. Since $x^2h = 12$, $h = \frac{12}{x^2}$, so $S = x^2 + \frac{48}{x}$. Finding S' and setting it equal to 0 yields $x = 2\sqrt[3]{3}$ as the only critical value. Since S' is negative to the left and positive to the right of $2\sqrt[3]{3}$, we have an absolute minimum.

Additional Practice

1. (A)

2. Maximum volume is approximately 52.5 in^3. Let each square measure x by x (inches) with $x \in [0, 4]$.

$$V(x) = x(8 - 2x)(10 - 2x) \Rightarrow V'(x) = 4(3x^2 - 18x + 20). \ V'(x) = 0 \text{ for } x = \frac{9 \pm \sqrt{21}}{3}.$$

$$V(x) \text{ is maximized at } x = \frac{9 - \sqrt{21}}{3} \text{ with volume } V\left(\frac{9 - \sqrt{21}}{3}\right).$$

3.

    7   Justify that your solution provides a maximum or minimum.

    5   Find the derivative of the varying quantities.

    4   If necessary, substitute from the fixed quantity into the varying quantity.

    8   Make sure you have answered the original question, with appropriate units of measure.

    2   Define variables to be used.

    1   Read and understand the problem, noting especially what is to be optimized.

    6   Find the zeros of the derivative equation.

    3   Write equations for all fixed and varying quantities.

Related Rates

Content & Practice

1. Solution given in text.

2. (A) Cube edge $= e$, $V_{\text{cube}} = e^3$ (B) $\dfrac{de}{dt} = 2$ in./min

    (C) Find dV/dt when $e = 3$ inches.   (D) $\dfrac{dV}{dt} = 3e^2 \dfrac{de}{dt}$

    (E) $\dfrac{dV}{dt} = 3(3 \text{ in.})^2 \left(2 \text{ in./min}\right)$ (F) $\dfrac{dV}{dt} = 54$ in.^{3}/min

3. (A) $y = (x - 3)^2$, $\dfrac{dx}{dt} = 4$ units/sec, find $\dfrac{dy}{dt}$ at $x = 1$.

    $\dfrac{dy}{dt} = 2(x - 3)\dfrac{dx}{dt} \Rightarrow$ at $x = 1$, $\dfrac{dy}{dt} = -16$ units/sec. This means that at the point $(1, 4)$

    the particle is moving downward along the parabolic path.

(B) The particle's distance from any position (x, y) to the origin is determined by the distance

formula $D = \sqrt{(x-0)^2 + (y-0)^2} = \sqrt{x^2 + y^2} = \sqrt{x^2 + (x-3)^4}$. The rate of

change of that distance will be $\dfrac{dD}{dt} = \dfrac{1}{2}\left(x^2 + y^2\right)^{-1/2}\left(2x\dfrac{dx}{dt} + 2y\dfrac{dy}{dt}\right)$. Evaluate at

$x = 1$, replacing $\dfrac{dx}{dt}, \dfrac{dy}{dt}$ with the values from part (A). $\dfrac{dD}{dt} = \dfrac{-60}{\sqrt{17}}$ units/sec.

4.   Ship W travels distance w and $dw/dt = 35$ mph

Ship H travels distance h and $dh/dt = 28$ mph

If the ships are distance a apart, find da/dt, when $t = 2$ hours.

$$a^2 = w^2 + h^2 - 2wh\cos\theta \Rightarrow 2a\frac{da}{dt} = 2w\frac{dw}{dt} + 2h\frac{dh}{dt} - 2\cos\theta\left(w\frac{dh}{dt} + h\frac{dw}{dt}\right)$$

Since $t = 2$ hours $\Rightarrow w = 70$ miles, $h = 56$ miles, $a = 54.74$ miles, $da/dt \approx 27.4$ mph.

5. $\dfrac{dr}{dt} = \dfrac{16}{25\pi}$ in./min. Given, $\dfrac{dV}{dt} = 4$ in.^{3}/min. We know that $V = \dfrac{1}{3}\pi r^2 h$ and from the similar

triangles (see diagram in text), $\dfrac{r}{h} = \dfrac{4}{16}$. Therefore, in terms of r, $V = \dfrac{4}{3}\pi r^3$. Find $\dfrac{dr}{dt}$ when

$h = 5$ in. $\dfrac{dV}{dt} = 4\pi r^2\dfrac{dr}{dt}$, when $h = 5$ in., $r = 1.25$ in., $\dfrac{dr}{dt} = \dfrac{16}{25\pi}$ in./min.

Additional Practice

1.   (E)   Let $y = $ length of diagonal, $\dfrac{y}{\sqrt{2}} = $ length of sides of square. $\dfrac{dy}{dt} = 3$ in./min. Find $\dfrac{dp}{dt}$

when $A = 18$ in.^{2} (therefore, $y = 6$ in.)

Perimeter:

$$p = 2\sqrt{2}y \Rightarrow \frac{dp}{dt} = 2\sqrt{2}\frac{dy}{dt}. \frac{dp}{dt} = 6\sqrt{2} \text{ in./min.}$$

2.   (C) $\dfrac{dV_{\text{sphere}}}{dt} = -2$ in.^{3}/min. If S is the surface area, find $\dfrac{dS}{dt}$ when $r = 1$ in. First,

$\dfrac{dV}{dt} = 4\pi r^2\dfrac{dr}{dt}$. Substituting at $r = 1$ gives $\dfrac{dr}{dt} = \dfrac{-1}{2\pi}$ in./min. $S = 4\pi r^2 \Rightarrow \dfrac{dS}{dt} = 2\pi r\dfrac{dr}{dt}$.

Substituting, $\dfrac{dS}{dt} = -1$ in.^{2}/min. $\Rightarrow$ decreases at 1 in.^{2}/min.

Implicit Differentiation

Content & Practice

1. $2x + 2y\dfrac{dy}{dx} = 6y^2\dfrac{dy}{dx} \Rightarrow \dfrac{dy}{dx} = \dfrac{2x}{6y^2 - 2y}$.

2. (A) $(2)(1)^2 + 2(1)^4 = 4$ and $2^2 \cdot 1 = 4$.

   (B) $x2y\dfrac{dy}{dx} + y^2 + 8y^3\dfrac{dy}{dx} = x^2\dfrac{dy}{dx} + 2xy \Rightarrow \dfrac{dy}{dx} = \dfrac{2xy - y^2}{2xy + 8y^3 - x^2}$. At $(2, 1)$, $\dfrac{dy}{dx} = \dfrac{3}{8}$.

Additional Practice

1. (C) $\dfrac{dy}{dx} = \dfrac{2x - y}{x + 2y}$. At $(2, 3)$, $\dfrac{dy}{dx} = \dfrac{1}{8}$.

2. (A) $\dfrac{dy}{dx} = \dfrac{2x - y^2}{2xy - 3y^2}$.

3. (A) $2x - \left(x^2\dfrac{dy}{dx} + 2xy\right) = 2y\dfrac{dy}{dx} \Rightarrow \dfrac{dy}{dx} = \dfrac{2x - 2xy}{x^2 + 2y}$.

   (B) Two points: $(2, -5)$ and $(2, 1)$. At $(2, 1)$, $\dfrac{dy}{dx} = 0 \Rightarrow$ a horizontal tangent.

4. $y = e^{\left(\frac{1}{2}x^2 + 5\right)} - 1$, $\dfrac{dy}{dx} = x \cdot e^{\left(\frac{1}{2}x^2 + 5\right)}$, $\dfrac{d^2y}{dx^2} = (x^2 + 1)e^{\left(\frac{1}{2}x^2 + 5\right)}$.

Derivative as a Rate of Change

Content & Practice

1. (A) $v(t) = s'(t) = 2t - 3 \Rightarrow v(5) = 2(5) - 3 = 7$ in./sec.

   (B) $v_{\text{avg}} = \dfrac{s(4) - s(0)}{4 - 0} = \dfrac{4 - 0}{4} = 1$ in./sec.

   (C) Speed $= |v(1)| = |(2 \cdot 1)^2 - 3| = |-1| = 1$ in./sec.

2. (A) Approximately $(0, 2) \cup (4.7, 6)$. Acceleration is the slope of the velocity graph, so it is positive when the velocity graph is rising.

(B) Speed is the absolute value of velocity (see the graph). This is decreasing on approximately $(2, 3) \cup (4.7, 6)$.

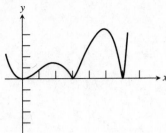

(C) $a_{\text{avg}} = \dfrac{v(5) - v(1)}{5 - 1} \approx \dfrac{-4.2 - 1}{4} = \dfrac{-5.2}{4} = -1.3 \ \text{ft/sec}^2.$

Additional Practice

1. (A) $v(t) = s'(t) = 1 + (-9)(t + 1)^{-2} = 1 - \dfrac{9}{(t + 1)^2} \Rightarrow v(1) = 1 - \dfrac{9}{2^2} = \dfrac{-5}{4}.$

2. (A) $v(15) \approx \dfrac{D(20) - D(10)}{20 - 10} = \dfrac{30 - 17}{10} = 1.3 \ \text{ft/sec}.$

    (B) $v_{\text{avg}} = \dfrac{D(20) - D(5)}{20 - 5} = \dfrac{30 - 7}{15} \approx 1.5 \ \text{ft/sec}.$

3. Acceleration is drawn in bold on the graph.

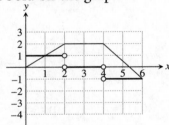

4. (C) Rate at which the volume of water in the tank is changing:
    $R'(t) = 6t^2 - 40t - 72$. Minimize this.

    $R''(t) = 12t - 40 = 0 \Rightarrow t = \dfrac{40}{12} = 3\frac{1}{3}$ min. R'' is changing from negative to positive at this value, so R' has a relative minimum there. Since $R'\left(3\frac{1}{3}\right) < 0$, the water is draining then, at its maximum rate.

Slope Fields

Content & Practice

1. (A)

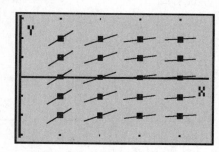

2. (D) A, B, and C would all have negative slopes on the left side. E has decreasing slopes toward the *y*-axis.

3. The slopes are zero whenever $x = 0$, positive in the first and third quadrants, and negative in the second and fourth quadrants.

4.

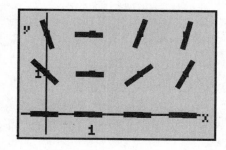

Additional Practice

1. (A)

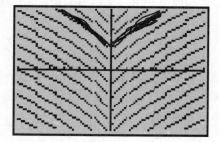

(B)

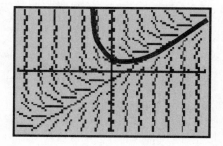

2. (A) The slope should be negative for $x < 0$, zero for $x = 0$, and positive for $x > 0$. The only graph with these properties is (A).

Euler's Method

Content & Practice

1.

(x, y)	$f'(x, y)$	Δx or h	$\Delta y = f'(x, y)\Delta x$	$(x + \Delta x, y + \Delta y)$
$(1, 1)$	3	0.5	1.5	$(1.5, 2.5)$
$(1.5, 2.5)$	4	0.5	2	$(2, 4.5)$

$$f'(1, 1) = 2(1) + 1 = 3 \Rightarrow \Delta y = 3(0.5) = 1.5 \Rightarrow (x + \Delta x, y + \Delta y) = (1 + 0.5, 1 + 1.5)$$
$$f'(1.5, 2.5) = 2(1.5) + 1 = 4 \Rightarrow \Delta y = 4(0.5) = 2 \Rightarrow (x + \Delta x, y + \Delta y) = (1.5 + 0.5, 2.5 + 2)$$

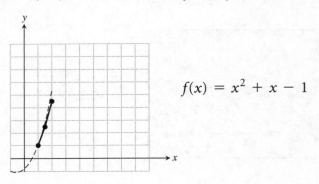

$$f(x) = x^2 + x - 1$$

2. Decreasing the value of Δx would provide a more accurate line. Or evaluate f' at the midpoint of the interval, rather than the left endpoint.

3.

(x, y)	$f'(x, y)$	Δx or h	$\Delta y = f'(x, y)\Delta x$	$(x + \Delta x, y + \Delta y)$
$(0, 1)$	1	0.5	0.5	$(0.5, 1.5)$
$(0.5, 1.5)$	2.5	0.5	1.25	$(1, 2.75)$
$(1, 2.75)$	4.75	0.5	2.375	$(1.5, 5.125)$

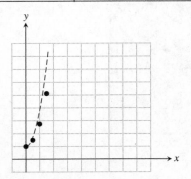

Additional Practice

1. **(C)** $f'(2, 1) = \frac{2}{1} = 2 \Rightarrow \Delta y = 2(0.5) = 1 \Rightarrow$

$$(x + \Delta x, y + \Delta y) = (2 + 0.5, 1 + 1) = (2.5, 2).$$

$f'(2.5, 2) = \frac{2.5}{2} = 1.25 \Rightarrow \Delta y = 1.25(0.5) = 0.625 \Rightarrow$

$$(x + \Delta x, y + \Delta y) = (2.5 + 0.5, 2 + 0.625) = (3, 2.625).$$

2. **(B)** $f'(1, 3) = 2(1) + 3 = 5 \Rightarrow \Delta y = 5(0.2) = 1 \Rightarrow$

$$(x + \Delta x, y + \Delta y) = (1 + 0.2, 3 + 1) = (1.2, 4).$$

$f'(1.2, 4) = 2(1.2) + 3 = 5.4 \Rightarrow \Delta y = 5.4(0.2) = 1.08 \Rightarrow$

$$(x + \Delta x, y + \Delta y) = (1.2 + 0.2, 4 + 1.08) = (1.4, 5.08).$$

$\frac{dy}{dx} = 2x + 3 \Rightarrow y = x^2 + 3x + c \Rightarrow 3 = 1^2 + 3(1) + c \Rightarrow$

$$3 = 4 + c \Rightarrow c = -1 \Rightarrow y = x^2 + 3x - 1 \Rightarrow$$

$y(1.4) = (1.4)^2 + 3(1.4) - 1 = 5.16$

error $= |5.08 - 5.16| = 0.08.$

L'Hôpital's Rule

Content & Practice

1. $\frac{1}{2}$

2. $\frac{-1}{3}$

3. 0

4. 5

5. 0. $\lim\limits_{x \to 3^+} (x - 3)^{1/(x-3)}$ is a limit of the form 0^{∞}. Thus the limit is zero.

6. 1. $\lim\limits_{x \to 0^+} \ln(\sin x)^x = \lim\limits_{x \to 0^+} [x \ln(\sin x)] = \lim\limits_{x \to 0^+} \frac{\ln(\sin x)}{x^{-1}} = \lim\limits_{x \to 0^+} x^2 \tan x = 0,$
   then $e^0 = 1.$

7. $e^{2/3}$ $\displaystyle\lim_{x\to\infty} \ln\left(1 + \frac{2}{3x}\right)^x = \cdots = \lim_{x\to\infty}\left(\frac{2}{3}\cdot\frac{1}{1 + \dfrac{2}{3x}}\right) = \frac{2}{3}$, then $e^{2/3}$.

Additional Practice

1. (B) $\displaystyle\lim_{x\to 0}\frac{\sqrt[3]{1 + x} - \frac{1}{3}x - 1}{x^2} = \cdots = \lim_{x\to 0}\left[-\frac{1}{9}(1 + x)^{-5/3}\right] = -\frac{1}{9}.$

2. (C) $\displaystyle\lim_{x\to\infty}\frac{\ln\sqrt{x}}{\ln(2 + 3x)} = \lim_{x\to\infty}\frac{2 + 3x}{6x} = \frac{1}{2}.$

3. (E) $\displaystyle\lim_{x\to 0}(e^x + 2x)^{1/x} = \lim_{x\to 0}\frac{e^x + 2}{e^x + 2x} = 3$, then e^3.

Basic Derivatives

Content & Practice

1. See text.

2. The derivative of a constant is 0.

3. $\dfrac{dy}{dx} = \cos x$

4. $\dfrac{dy}{dx} = -\sin x$

5. $\dfrac{dy}{dx} = \sec^2 x$

6. $y = \cot x$

7. $\dfrac{dy}{dx} = \sec x \tan x$

8. $y - \csc x$

9. $\dfrac{dy}{dx} = \dfrac{1}{x}$

10. $y = e^x$

11. $\dfrac{dP}{dw} = \sec^2 w$

12. $\dfrac{dV}{dr} = \dfrac{2}{3}\pi rh$

13. $\dfrac{dE}{dm} = c^2$

14. $\dfrac{dS}{dt} = 12t$

Additional Practice

1. (D) $\dfrac{dy}{dx} = \cos x$

   $\cos\dfrac{\pi}{3} = \dfrac{1}{2}$

2. (D) $\dfrac{dy}{dx} = e^x$

   At $x = a$, $\dfrac{dy}{dx} = e^a$.

3. (B) $\dfrac{dA}{dr} = 2\pi r$

At $r = 2$, $\dfrac{dA}{dr} = 4\pi$ (Instantaneous)

$\dfrac{A(3) - A(1)}{3 - 1} = \dfrac{9\pi - \pi}{2} = 4\pi$ (Average)

4. $\dfrac{1}{3}x = \sqrt{x}$

$\dfrac{1}{9}x^2 = x$

$x^2 - 9x = 0$

$x = 9 \Rightarrow y = 3$

Slope is $\dfrac{dy}{dx} = \dfrac{1}{2}x^{-1/2}$.

At $x = 9$, slope is $\dfrac{1}{6}$.

$y - 3 = \dfrac{1}{6}(x - 9)$

Derivative Rules

Content & Practice

Formula Name	Derivative Formula	Word Description
Difference Formula	$y = f(x) - g(x)$ $\dfrac{dy}{dx} = f'(x) - g'(x)$	The derivative of a difference of functions is the difference of their individual derivatives.
Sum Formula	$y = f(x) + g(x)$ $\dfrac{dy}{dx} = f'(x) + g'(x)$	The derivative of a sum of functions is the sum of the derivative of each function.
Product Rule	$y = f(x) \cdot g(x)$ $\dfrac{dy}{dx} = f'(x) \cdot g(x) + f(x) \cdot g'(x)$	The derivative of a product of functions is the derivative of the first function times the second function, plus the derivative of the second function times the first function.
Quotient Rule	$y = \dfrac{f(x)}{g(x)}$ $\dfrac{dy}{dx} = \dfrac{g(x) \cdot f'(x) - f(x) \cdot g'(x)}{[g(x)]^2}$	The derivative of a quotient of functions is the denominator times the derivative of the numerator, minus the numerator times the derivative of the denominator, all over the square of the denominator.

1. $y = 5x^3 + \ln x \qquad \dfrac{dy}{dx} = 15x^2 + \dfrac{1}{x}$

2. $y = x^3 \cdot \ln(x) \qquad \dfrac{dy}{dx} = x^3 \cdot \dfrac{1}{x} + \ln x \cdot 3x^2 = x^2 + 3x^2 \ln x$

3. $s = \dfrac{5}{t^4} \qquad \dfrac{ds}{dt} = -20t^{-5} = \dfrac{-20}{t^5}$ \quad Quotient rule is not required.

4. $k = \dfrac{\sin p}{\sqrt{p}} \qquad \dfrac{dk}{dp} = \dfrac{\sqrt{p}\cos p - \dfrac{1}{2\sqrt{p}}\sin p}{(\sqrt{p})^2} = \dfrac{2p\cos p - \sin p}{2p^{2/3}}$

5. $w = z^4 - (z+2)^7 \cos z \qquad \dfrac{dw}{dz} = 4z^3 - [(z+2)^7 - \sin z + \cos z\, 7(z+2)^6]$

$$= 4z^3 + \sin z(z+2)^7 - 7\cos z(z+2)^6$$

6. $y = x^2(x^3 - 2) \qquad y = x^5 - 2x^2 \Rightarrow \dfrac{dy}{dx} = 5x^4 - 4x$

$$\dfrac{dy}{dx} = x^2 3x^2 + (x^3 - 2)2x$$

$$= 3x^4 + 2x^4 - 4x$$

$$= 5x^4 - 4x$$

Additional Practice

1. (B) $\quad f'(x) = e^x - \sin x + \cos x\, e^x = 0$

$e^x(-\sin x + \cos x) = 0$

$\sin x = \cos x$ and $e^x \neq 0$

$x = \dfrac{\pi}{4}$

2. (E) $\quad y = \dfrac{x+3}{x^2+1}\bigg|_{x=1} = 2 \qquad$ Ordered pair is $(1, 2)$.

$$\dfrac{dy}{dx} = \dfrac{(x^2+1)1 - (x+3)2x}{(x^2+1)^2}$$

At $x = 1, \dfrac{dy}{dx} = \dfrac{2-8}{4} = -\dfrac{3}{2}.$

$$y - 2 = -\dfrac{3}{2}(x-1)$$

$$y = -\dfrac{3}{2}x + \dfrac{7}{2}$$

3. (C) $\dfrac{d}{dx}(f \cdot g)\bigg|_{x=3} = f(3) \cdot g'(3) + g(3) \cdot f'(3)$

$$= 7 \cdot -1 + -4 \cdot \dfrac{3}{2}$$

$$= -13$$

4. (A) $\dfrac{d}{dx}\left(\dfrac{f}{g}\right)\bigg|_{x=1} = \dfrac{g(1) \cdot f'(1) - f(1) \cdot g'(1)}{[g(1)]^2}$

$$= \dfrac{2 \cdot 5 - 4 \cdot \dfrac{1}{2}}{2^2}$$

$$= 2$$

Chain Rule

Content & Practice

1. $\dfrac{dy}{dx} = \dfrac{dy}{du} \cdot \dfrac{du}{dx} = (3u^2)(4) = 12u^2 = 12(4x + 2)^2$

2. Let $g(v) = 1 - \sin v, h(g) = g^{1/2}$.

$$\dfrac{dh}{dv} = \dfrac{dh}{dg} \cdot \dfrac{dg}{dv} = \left(\dfrac{1}{2}g^{-1/2}\right)\left(-\cos v\right) = -\dfrac{1}{2}\left(1 - \sin v\right)^{-1/2}\cos v$$

3. $y = [\sin(t^2 + 5)]^3$. Let $u = t^2 + 5, v = \sin u, y = v^3$.

$$\dfrac{dy}{dt} = \dfrac{dy}{dv} \cdot \dfrac{dv}{du} \cdot \dfrac{du}{dt} = (3v^2)(\cos u)(2t) = 6\sin^2(t^2 + 5)\cos(t^2 + 5)t$$

4. (D)

5. (A) $\dfrac{dy}{dx} = 0$ (B) $\dfrac{dy}{dx} = -\sin u \cdot \dfrac{du}{dx}.$

(C) $\dfrac{dy}{dx} = \sec^2 u \cdot \dfrac{du}{dx}.$ (D) $y = \cot u$

(E) $\dfrac{dy}{dx} = \sec u \cdot \tan u \cdot \dfrac{du}{dx}.$ (F) $y = \csc u$

(G) $\dfrac{dy}{dx} = e^u \cdot \dfrac{du}{dx}$

(H) $y = \ln u$

(I) $\dfrac{dy}{dx} = \dfrac{a^u}{\ln a} \cdot \dfrac{du}{dx}$

(J) $\dfrac{dy}{dx} = \dfrac{1}{\sqrt{1 - u^2}} \cdot \dfrac{du}{dx}, |u| < 1$

(K) $y = \cot^{-1} u$

(L) $\dfrac{dy}{dx} = \dfrac{1}{|u|\sqrt{u^2 - 1}} \cdot \dfrac{du}{dx}, |u| > 1$

(M) $y = \csc^{-1} u$

(N) $\dfrac{dy}{dx} = \dfrac{1}{1 + u^2} \cdot \dfrac{du}{dx}$

(O) $\dfrac{dy}{dx} = \dfrac{-1}{\sqrt{1 - u^2}} \cdot \dfrac{du}{dx}, |u| < 1$

(P) $\dfrac{dy}{dx} = \dfrac{1}{u \cdot \ln a} \cdot \dfrac{du}{dx}$,

Additional Practice

1. (C) $y = \left[f(x)\right]^2 \Rightarrow y' = 2f(x)f'(x)$. $y'(2) = 2f(2)f'(2) = 2(7)\left(\dfrac{1}{2}\right) = \dfrac{14}{3}$

2. (C) $y' = f'(g(x))g'(x) y'(1) = f'(g(1))g'(1) = f'(2)g'(1) = \dfrac{1}{4}$

Derivatives of Parametric, Polar, and Vector Functions

Content & Practice

1. Solutions in text.

2. $\dfrac{dy}{dx} = -\dfrac{\cos 2t}{2 \sin t}, \dfrac{d^2y}{dx^2} = -\dfrac{2 \sin t \sin 2t + \cos 2t \cos t}{8 \sin^3 t}$. At $t = \dfrac{\pi}{2}, \dfrac{d^2y}{dx^2} = 0$.

3. $f''(t) = (4e^{2t}, -\cos t)$

Additional Practice

1. $y = -2.229x + 5.229. \dfrac{dy/dt}{dx/dt} = \dfrac{-5}{3 + \sin 4} = -2.229$.

2. (C) $\dfrac{dy}{dx} = \dfrac{3}{2(2t - 1)\sqrt{3t + 1}}$. At $t = 1, \dfrac{dy}{dx} = \dfrac{3}{4}$.

3. (A) $f''(1) = -i - \dfrac{1}{4}j. f''(t) = -\dfrac{1}{t^2}i - \dfrac{1}{4}t^{-3/2}j$.

Integrals

Riemann Sums

Content & Practice

1. (A) 10. Left-hand rectangles and four equal subdivisions: $[0 + 3 + 4 + 3] \cdot 1 = 10$

    (B) 10. Right-hand rectangles and four equal subdivisions: $[3 + 4 + 3 + 0] \cdot 1 = 10$

    (C) 11. Midpoint rectangles and four equal subdivisions:
    $[1.75 + 3.75 + 3.75 + 1.75] \cdot 1 = 11$

2. (A) Left-hand rectangles:
    $[0 + 6 + 10 + 16 + 14 + 12 + 18 + 22 + 12 + 4] \cdot 1 = 114$ inches

    (B) Right-hand rectangles:
    $[6 + 10 + 16 + 14 + 12 + 18 + 22 + 12 + 4 + 2] \cdot 1 = 116$ inches

Additional Practice

1. (B) $[4.3 \cdot 1.2 + 3.1 \cdot 1.1 + 2.2 \cdot 1.5 + 1.5 \cdot 1.6] = 14.27$

2. 93°C Left-hand Riemann sum, subintervals of length 5 minutes.

$$\frac{1}{25 - 0} [24 + 76 + 106 + 124 + 135]5 = 93°C$$

Definite Integral of a Rate of Change

Content & Practice

1. (A) $6\frac{1}{3}$ cm. Net change in position $= \int_0^2 4 - t^2 \, dt = 5\frac{1}{3}$. Given the $x = 1$ starting point,

    the particle position after 2 seconds is $6\frac{1}{3}$.

    (B) 4 cm. The net change in position is 3 cm. The position after 3 seconds is
    $x = 1 + 3 = 4$.

2. 34,925 bushels. $\int_0^{14} 1.3 + 1.025^t \, dt = 34.925$ thousand bushels

Additional Practice

1. (D) $\int_0^{24} 25 e^{-0.25(t-15)^2} \, dt = 197.728$

2. (A) 1284 bison.   Herd growth from 2000 to 2015 is approximated by
$\int_0^{15} 26.7 \cdot 1.036^t \, dt = 528$. Therefore, the herd size at the beginning of 2015 will be
approximately $756 + 528 = 1284$ bison.

(B) 35 bison per year. $\dfrac{528}{15} \, dt = 35.220$ bison per year

Basic Properties of Definite Integrals

Content & Practice

1. (A) 0.   See Rule 2.
   (B) 8. $3 \int_1^2 f(x) \, dx + \int_1^2 1 \, dx = -9 + 1 = -8$. See Rule 4.
   (C) 8. $\int_1^5 f(x) \, dx - \int_1^2 f(x) \, dx = 5 - (-3) = 8$. See Rule 5.

2. 8. $4(3 - 1) = 8$. See Rule 6.

3. $-\pi$. $g(-2) = \int_0^{-2} f(t) \, dt = -\int_{-2}^0 f(t) \, dt = \dfrac{-\pi 2^2}{4} = -\pi$. $\int_{-2}^0 f(t) \, dt$ represents one-fourth
of the circle's area; given radius 2 and the familiar area formula $\frac{1}{4}\pi r^2$, the value of the integral
is $\dfrac{\pi 2^2}{4}$ in the calculation.

Additional Practice

1. (E) $f(a)$ is a minimum and $f(b)$ is a maximum, so I and II are true.

2. (A) $\int_{-1}^3 h(r) \, dr - \int_{-1}^1 h(r) \, dr = 7 - 0 = 7$. See Rule 5.
   (B) $\int_1^3 h(r) \, dr = 7$. See Rule 1.

Applications of Integrals

Content & Practice

1. $10\frac{2}{3}$.   The area can be approximated by rectangular slices. The Riemann sum for these slices
is $\sum_{k=1}^n (4 - x^2)\Delta x$. The limit of this sum gives the definite integral $\int_{-2}^2 (4 - x^2) \, dx = 10\frac{2}{3}$.

2. $\int_0^{0.739}(\cos x - x) \, dx = 0.400$

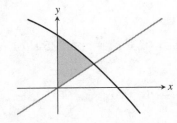

3. $\int_0^4 \left[\pi \left(\sqrt{x} \right)^2 - \pi \left(\frac{1}{2} x \right)^2 \right] dx = \frac{8\pi}{3}$

4. $\int_0^1 \sqrt{(2t)^2 + (3t^2)^2} \, dt = \int_0^1 t\sqrt{4 + 9t^2} \, dt$

Additional Practice

1. $\int_0^1 \frac{1}{2} \pi \left(\frac{1-x}{2} \right)^2 dx = \frac{\pi}{24}$

2. (A) $\frac{1}{2-0} \int_0^2 3x^2 \sqrt{x^3 + 1} \, dx = \frac{26}{3}$

3. $\frac{4}{\pi}$ units. Since the function is negative on $[1, 2]$, the particle is reversing direction and the corresponding integral must be subtracted to find total distance: $2 \int_0^1 \sin \pi t \, dt = \frac{4}{\pi}$.

4. $2\sqrt{3} + \frac{4}{3} \pi$. Sketch the graph and identify the region.

$\int_{\pi/3}^{-\pi/3} \frac{1}{2} (4 \cos \theta)^2 - \frac{1}{2} (2)^2 \, d\theta = 2\sqrt{3} + \frac{4}{3} \pi$

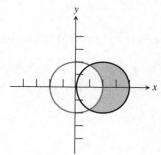

5. (C) Use the length of curve formula, $y'(x) = x^{1/2}$: $\int_0^3 \sqrt{1 + (\sqrt{x})^2} \, dx = \frac{14}{3}$.

Fundamental Theorem of Calculus

Content & Practice

1. $\frac{e^2 - 5}{2}$. $\int_1^e \left(x - \frac{2}{x} \right) dx = \frac{1}{2} x^2 - 2 \ln x \Big]_1^e = \frac{e^2 - 5}{2}$

2. $\sin x^2 \, 2x$. $\frac{d}{dx} \int_0^{x^2} \sin t \, dt = \frac{d}{dx} (\sin x^2) = \sin x^2 \, 2x$. (The $2x$ factor comes from the Chain Rule.)

3. (A) Maximum value occurs when area is maximum at $x = 1$. $f'(x) = f(x)$. From the graph, f is positive to the left of $x = 1$, zero at $x = 1$, and negative to the right of $x = 1$.

(B) Minimum value at $x = 3$. Think of F as an area accumulation function. As x increases from zero, F accumulates area and increases until $x = 1$. To the right of $x = 1$, F loses value until $x = 3$. Since F lost more than it gained, it has a minimum value at $x = 3$.

(C)

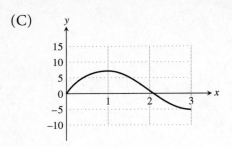

Additional Practice

1. -2^{x^2}

2. (A) $g(-1) = \int_1^{-1} f(t)dt = -\int_{-1}^1 f(t)dt = -\left(\frac{1}{2}\right)(2)(1) = -1.$

    $g(1) = \int_1^1 f(t)dt = 0.$

    $g(3) = \int_1^3 f(t)dt - \frac{1}{2} \cdot \pi(1)^2 = \frac{\pi}{2}.$

    (B) Never. g does not decrease because $g(x) = f(x)$ is never negative.

    (C) The graph of g is concave downward on the open intervals $(-1, 1)$ and $(2, 3)$ because $g''(x) = f'(x)$ would be negative on those intervals.

3. (B) $\int_0^x (t^2 - 3t)\, dt = \frac{x^3}{3} - \frac{3x^2}{2}, \int_2^x t\, dt = \frac{x^2}{2} - 2,$ $\frac{x^3}{3} - \frac{3x^2}{2} \geq \frac{x^2}{2} - 2$ for $x \leq 1.1$

4. (A) 6. $s(2) = \int_0^2 f(x)\, dx = \frac{1}{2}(2 + 4) \cdot 2 = 6$

    (B) $t = 7$. Velocity $= s'(t) = f(t)$. From the graph of f, the velocity is positive (particle moves to the right) until $t = 7$. Then the particle stops at $t = 7$ (velocity $= 0$). After $t = 7$, the velocity is negative so the particle moves to the left.

Antiderivative Basics

Content & Practice

1. (A) $\cos x$
    (B) $\sin x$
    (C) $\sec^2 x$
    (D) $\cot x + C$
    (E) $\sec x \tan x$
    (F) $\csc x + C$
    (G) $\ln x + C, \quad (x > 0)$
    (H) e^x
    (I) $\log_a x, \quad (a > 0)$

2. Function: $\sin 2x$ Antiderivative: $-\dfrac{1}{2}\cos 2x + C$

3. Function: $8x^3 + \sqrt{x}$ Antiderivative: $2x^4 + \dfrac{2}{3}x^{3/2} + C$

4. Function: $3 + e^{5t}$ Antiderivative: $3t + \dfrac{1}{5}e^{5t} + C$

5. Function: $x \cos x^2$ Antiderivative: $\dfrac{1}{2}\sin x^2 + C$

6. Function: $8x - \csc x \cot x$ Antiderivative: $4x^2 + \csc x + C$

7. Function: $\sec^2 5x$ Antiderivative: $\dfrac{1}{5}\tan 5x + C$

8. Function: $6(2x + 7)^5$ Antiderivative: $\dfrac{1}{2}(2x + 7)^6 + C$

9. Function: $3^{4x}\ln(3)$ Antiderivative: $\dfrac{1}{4}3^{4x} + C$

10. A function $F(x)$ is an antiderivative of a function $f(x)$ of $F'(x) = f(x)$ for all x in the domain of f.

11. Functions differing by a constant have the same derivative.

Additional Practice

1. (B) $x^2 + 7x + C$

2. (D) $\displaystyle\int_0^{\pi/6}\sin 2x\, dx = -\dfrac{1}{2}\cos 2x \bigg]_0^{\pi/6} = -\dfrac{1}{2}\left(\cos\dfrac{\pi}{3} - \cos 0\right) = \dfrac{1}{4}.$

3. (A) $f(x) = 3 + \ln(x^2 + 1)$ is a possibility because the antiderivative of $f'(x)$ is $\ln(x^2 + 1) + C.$

Integration by Substitution

Content & Practice

1. $u = x^3$ and $du = 3x^2\, dx$, substitution transforms the original integral to $\int u\, du.$

2. $u = 4x + 3$ and $du = 4\, dx$, substitution transforms the original integral to $\dfrac{5}{4}\int\dfrac{1}{u}\, du.$

3. $u = \tan x$ and $du = \sec^2 x \, dx$, substitution transforms the original integral to $\int 7u^5 \, du$.

4. $u = x^2 - 4$ and $du = 2x \, dx$, substitution transforms the original integral to $2 \int u^{1/2} \, du$.

5. $u = x^2$ and $du = 2x \, dx$, substitution transforms the original integral to $4 \int \csc u \cot u \, du$.

6. $u = \ln x$ and $du = \frac{1}{x} \, dx$, substitution transforms the original integral to $\int 3^u \, du$.

7. $u = x^3 + 5x$ and $du = (3x^2 + 5) \, dx$, substitution transforms the original integral to $2 \int \frac{du}{u}$. From there, $2 \int \frac{du}{u} = 2 \ln u + C$. Substituting back gives

$$\int \frac{6x^2 + 10}{x^3 + 5x} \, dx = 2 \ln(x^3 + 5x) + C.$$

Additional Practice

1. (C) $u = \cos x$ and $du = -\sin x \, dx$; the original integral becomes $-\int e^u \, du$, which integrates to (C).

2. (E) $u = \sin 2x$ and $du = 2 \cos 2x \, dx$; the original integral becomes $\frac{1}{2} \int_0^{\sqrt{3}/2} u \, du$. Note the changes in the limits of integration.

3. (B) $u = x^4 + 2x$ and $du = (4x^3 + 2) \, dx$; the original integral becomes $\frac{1}{2} \int_3^{20} \frac{du}{u}$. Note the changes in the limits of integration. Integration yields

$$\frac{1}{2} \left(\ln 20 - \ln 3 \right) = \frac{1}{2} \ln \frac{20}{3} = \ln \left(\frac{20}{3} \right)^{1/2} \text{ or } \ln \sqrt{\frac{20}{3}}.$$

4. (D) Let $u = 2x$ and $du = 2 \, dx$, then $\int_2^4 f(2x) \, dx$ becomes $\frac{1}{2} \int_4^8 f(u) \, du$. Note the changes in the limits of integration. So, $\int_2^4 f(2x) \, dx = \frac{1}{2} \int_4^8 f(u) \, du$. Therefore,

$$\int_4^8 f(u) \, du = 2 \int_2^4 f(2x) \, dx = 2(10) = 20.$$

Antidifferentiation by Parts

Content & Practice

1. Let $u = \ln x$ and $dv = x^2 \, dx$.

2. Let $u = \tan^{-1} x$ and $dv = x \, dx$.

3. Let $u = x$ and $dv = \cos x\, dx$.

4. Direct substitution; let $u = \sqrt{x} = x^{1/2}$ and $du = \frac{1}{2} x^{-1/2}\, dx$.

5. Let $u = x$ and $dv = 5^x\, dx$.

6. Let $u = \ln x^2$ and $dv = x\, dx$, so $du = \frac{2}{x}\, dx$ and $v = \frac{1}{2} x^2$.

$$\int x \ln x^2\, dx = \frac{1}{2} x^2 \ln x^2 - \int \left(\frac{1}{2} x^2\right) \frac{2}{x}\, dx = x^2 \ln x - \int x\, dx = x^2 \ln x - \frac{1}{2} x^2 + C$$

Additional Practice

1. (C) Let $u = \ln x$ and $dv = x^{-2}\, dx$, so $du = \frac{1}{x}\, dx$ and $v = -x^{-1}$.

2. (D) Let $u = \sin^{-1} x$ and $dv = dx$, so $du = \dfrac{1}{\sqrt{1 - x^2}}\, dx$ and $v = x$.

$$\int \sin^{-1} x\, dx = x \sin^{-1} x - \int x \frac{1}{\sqrt{1 - x^2}}\, dx = x \sin^{-1} x + \sqrt{1 - x^2} + C$$

3. (A) Area $= \int_0^\pi [x \sin x - (x - \pi)]\, dx = \int_0^\pi x \sin x\, dx - \int_0^\pi (x - \pi)\, dx$. Integrate the first integral by parts, let $u = x$ and $dv = \sin x\, dx$.

Antidifferentiation by Simple Partial Fractions

Content & Practice

1. $\dfrac{8}{2x^2 - 3x - 2} = \dfrac{8}{(2x + 1)(x - 2)} = \dfrac{A}{2x + 1} + \dfrac{B}{x - 2}$. Let $x = 2$ and the equation becomes $8 = 5B$. Let $x = -\dfrac{1}{2}$ and the equation becomes $8 = -\dfrac{5}{2} A$. Thus,

$$\frac{8}{2x^2 - 3x - 2} = \frac{-\frac{16}{5}}{2x + 1} + \frac{\frac{8}{5}}{x - 2}.$$

2. $\dfrac{7x - 1}{x^2 + 4x - 21} = \dfrac{7x - 1}{(x + 7)(x - 3)} = \dfrac{A}{x + 7} + \dfrac{B}{x - 3}$. Let $x = 3$ and the equation becomes $B = 2$. Let $x = -7$ and the equation becomes $A = 5$. Thus,

$$\frac{7x - 1}{x^2 + 4x - 21} = \frac{5}{x + 7} + \frac{2}{x - 3}.$$

Additional Practice

1. (B) First use simple partial fractions: $\dfrac{4}{x^2 + 8x + 15} = \dfrac{4}{(x + 3)(x + 5)} = \dfrac{2}{x + 3} + \dfrac{-2}{x + 5}.$

Then integrate:

$$\int \frac{4}{x^2 + 8x + 15} = \int \frac{2}{x + 3}\,dx - \int \frac{2}{x + 5}\,dx = 2\ln|x + 3| - 2\ln|x + 5| + C.$$

2. (E) First use simple partial fractions: $\dfrac{3x + 1}{x(x + 1)} = \dfrac{1}{x} + \dfrac{2}{x + 1}.$ Then integrate:

$$\int \frac{3x + 1}{x(x + 1)}\,dx = \int \frac{1}{x}\,dx + \int \frac{2}{x + 1}\,dx = \ln|x| + 2\ln|x + 1| + C = \ln\left|x(x + 1)^2\right| + C.$$

Given that $y(1) = \ln 8$, $C = \ln 2$, $\ln\left|x \cdot (x + 1)^2\right| + \ln 2 = \ln\left|2x(x + 1)^2\right|.$

Improper Integrals

Content & Practice

1. $\displaystyle\lim_{a\to\infty} \int_0^a \tan^{-1} x\,dx$

2. $\displaystyle\lim_{a\to 0} \int_a^5 \frac{1}{e^x - 1}\,dx$

3. $\displaystyle\lim_{a\to 2^-} \int_0^a \frac{dx}{\sqrt[3]{4 - x^2}} + \lim_{a\to 2^+} \int_a^6 \frac{dx}{\sqrt[3]{4 - x^2}}$

4. $\displaystyle\int_1^\infty \frac{dx}{1 + x^3}$ converges because $0 \le \dfrac{1}{1 + x^3} \le \dfrac{1}{x^3}$ on $[1, \infty)$ and $\displaystyle\int_1^\infty \frac{1}{x^3}\,dx$ converges.

5. $\displaystyle\int_1^\infty \frac{dx}{x^2 - 0.9}$ converges because the functions $f(x) = \dfrac{1}{x^2 - 0.9}$ and $g(x) = \dfrac{1}{x^2}$ are positive

and continuous on $[1, \infty)$. Also, $\displaystyle\lim_{x\to\infty} \frac{f(x)}{g(x)} = \lim_{x\to\infty} \frac{1/x^2}{1/(x^2 - 0.9)} = \lim_{x\to\infty} \frac{2x}{2x} = 1$ (L'Hôpital's

Rule). Thus $\displaystyle\int_1^\infty \frac{dx}{x^2 - 0.9}$ converges because $\displaystyle\int_1^\infty \frac{dx}{x^2}$ converges.

6. $\displaystyle\int_0^1 \frac{dx}{1 + \sqrt{x}}$ converges because $0 \le \dfrac{1}{1 + \sqrt{x}} \le \dfrac{1}{\sqrt{x}}$ on $[0, 1]$ and $\displaystyle\int_0^1 \frac{1}{\sqrt{x}}\,dx$ converges.

7. $\displaystyle\int_2^\infty \frac{dx}{x}$ diverges because $0 \le \dfrac{1}{x} \le \dfrac{1}{\ln x}$ on $[2, \infty]$ and $\displaystyle\int_2^\infty \frac{dx}{x}$ diverges.

Additional Practice

1. (C) $\int_0^\infty \frac{3}{2} e^{-x/2}\, dx = \frac{3}{2} \lim_{a\to\infty} \int_0^a e^{-x/2}\, dx = \frac{3}{2} \lim_{a\to\infty} \left(-2e^{-x/2}\right)\Big]_0^a = 3$

2. (D) $\int_2^\infty \dfrac{dx}{x^2 + 5x + 6} = \int_2^\infty \dfrac{dx}{(x+2)(x+3)} =$

   $\int_2^\infty \dfrac{1}{x+2} - \dfrac{1}{x+3}\, dx = \lim_{a\to\infty} \left(\ln|x+2| - \ln|x+3|\right)\Big]_2^a = \lim_{a\to\infty} \left(\ln\dfrac{a+2}{a+3} - \ln\dfrac{4}{5}\right) = \ln\dfrac{5}{4}$

3. (C) Applying integration by parts to the improper integral gives

   $$\int_0^1 x \ln x\, dx = \lim_{a\to 0^+} \left[\frac{1}{2}x^2 \ln x - \frac{1}{4}x^2\right)\Big]_a^1 = \lim_{a\to 0^+}\left[-\frac{1}{4} - \left(\frac{1}{2}a^2 \ln a - \frac{1}{4}a^2\right)\right].$$

   The trick here is to evaluate the expression $\displaystyle\lim_{a\to 0^+} a^2 \ln a = \lim_{a\to 0^+} \dfrac{\ln a}{a^{-2}} = 0$ using L'Hôpital's Rule.

Initial Value Problems

Content & Practice

1. Solution printed in text.

2. (A) $v(t) = -32t + 155$

   (B) $h(t) = -16t^2 + 155t + 5$

Additional Practice

1. (A) $a(t) = t + \cos t$, $v(t) = \frac{1}{2}t^2 + \sin t + C$. $v(0) = -3 \Rightarrow C = -3$. Setting $v = 0$,

   $\frac{1}{2}t^2 + \sin t - 3 = 0$, and solving for t, $t = 2.057$.

2. $x(t) = -\cos t + 2$, $a = \cos t \Rightarrow v = \sin t + C_1$ and given $v(0) = 0$, $C_1 = 0$. Then $v = \sin t \Rightarrow x = -\cos t + C_2$ and since $x(0) = 1$, $C_2 = 2$.

3. $f(4) \approx 0.3757$

Separable Differential Equations

Content & Practice

1. (A) $y = 2000e^{2.3t}$

   (B) $1.964 \cdot 10^{10}$

2. (A) $\dfrac{dP}{dt} = \dfrac{1}{36} P(12 - P)$

(B) $\lim\limits_{t \to \infty} P(t) = 12$

Additional Practice

1. $y(t) = 2e^{\left(\frac{t}{3} - \frac{t^2}{24}\right)}$. $\int \dfrac{dy}{y} = \int \left(\dfrac{1}{3} \cdot -\dfrac{t}{12}\right) dt \Rightarrow y = C \cdot e^{\left(\frac{t}{3} - \frac{t^2}{24}\right)}$. Since $y(0) = 2$, $C = 2$.

2. (A) Solving $y = y_0 e^{kt}$ using the given information gives

$2y_0 = y_0 e^{8k} \Rightarrow k = \dfrac{\ln 2}{8} \approx .0866434$.

3. $y = 4e^{1-x^2}$. $\int \dfrac{dy}{y} = \int -2x \, dx$. Integrating both sides and solving for y gives $y = Ce^{-x^2}$.
Since $y(1) = 4$, $C = 4e$ and $4e \cdot e^{-x^2} = 4e^{1-x^2}$.

4. (A) $\dfrac{dP}{dt} = 0.012P(500 - P)$ so by comparison to $\dfrac{dP}{dt} = kP(M - P)$,
carrying capacity is $M = 500$.

Numerical Approximation to Definite Integrals

Content & Practice

1. (A) 10. $\int_{-2}^{2} f(x) \, dx \approx \dfrac{1}{2} [f(-2) + 2f(-1) + 2f(0) + 2f(1) + f(2)] = 10$

(B) Left Riemann sum $= 10$; right $= 10$, average $= 10$

2. 115 ft. From the data: $\dfrac{1}{2} [v(0) + 2v(1) + 2v(2) + \cdots + 2v(8) + 2v(9) + v(10)] = 115$

Additional Practice

1. 6. From the graph: $\dfrac{1}{2} \left[f(-2) + 2f(-1) + 2f(0) + 2f(1) + f(2)\right] = 6$

2. (D) $\int_{1}^{7} f(x) \, dx \approx \left[\dfrac{1}{2} \cdot 3 \cdot (10 + 20) + \dfrac{1}{2} \cdot 2 \cdot (20 + 40) + \dfrac{1}{2} \cdot 1 \cdot (40 + 30)\right] = 140$

Polynomial Approximation

Concept of Series

Content & Practice

1. Converges to 1

2. (A) about 1.64

   (B) Diverges

Additional Practice

1. Converges

2. (C)

Geometric, Harmonic, and Alternating Series

Content & Practice

1. (B)

2. The error is less than the first unused term (1/5!), which is less that 1/100. Since the first unused term is positive, the error is positive and the approximation is an underestimate.

3. 15. $\dfrac{5}{1 - 2/3} = 15$

Additional Practice

1. (C) $S_\infty = \dfrac{\frac{3}{2}}{1 - \frac{3}{8}} = \dfrac{12}{5}$

2. (A) 3. $S_\infty = \dfrac{4}{1 + \frac{1}{3}} = 3$

   (B) 8 terms

Integral Test, Ratio Test, and Comparison Test

Content & Practice

1. $\sum_{n=1}^{\infty} \frac{1}{n^2}$ converges. Series of the form $\sum_{n=1}^{\infty} \frac{1}{n^p}$ converge if $p > 1$ and diverge if $p \leq 1$.

2. Converges by the Ratio Test ($L = 0$)

3. Converges by comparison with $\sum_{n=1}^{\infty} \frac{1}{n^2}$, which converges as a p-series with $p > 1$.

Additional Practice

1. (D)

Taylor Polynomials

Content & Practice

1. (A) At $x = 2$
   (B) Yes
   (C) Domain: $(1, 3]$

2. (A) (E). $-\frac{1}{2} < x < 2\frac{1}{2}$

   (B) (D). $|x| < 0.8$

   (C) (B). $|x| < 2\frac{1}{2}$

   (D) (A). $|x| < 2$

   (E) (C). $|x| < 1$

Additional Practice

1. (C)

2. (A) $\cos x = 1 - \frac{x^2}{2!} + \frac{x^4}{4!} - \frac{x^6}{6!} + \cdots$. Therefore,

$$\cos 2x = 1 - \frac{(2x)^2}{2!} + \frac{(2x)^4}{4!} - \frac{(2x)^6}{6!} + \cdots = P(x)$$

Maclaurin and Taylor Series

Additional Practice

1. $4 + 7x - x^2 + 5\dfrac{x^3}{3!} - 8\dfrac{x^4}{4!}$

2. $\sum_{k=0}^{n}(-1)^k(x-1)^k = 1 - (x-1) +$
    $$(x-1)^2 - (x-1)^3 + (x-1)^4 - \cdots + (-1)^n(x-1)^n + \cdots$$

3. (B) Set $\dfrac{h^{12}(0)}{12!} = \dfrac{1}{3960}$ and solve for $h^{12}(0)$.

Manipulating Taylor Series

Content & Practice

1. $\tan^{-1}x + C = C + x - \dfrac{x^3}{3} + \dfrac{x^5}{5} - \dfrac{x^7}{7} + \cdots = C + \sum_{n=1}^{\infty}\dfrac{x^{2n-1}}{2n-1}(-1)^{n+1}$; center at 0.

2. No. Substitution would produce a series that is centered at $\frac{1}{2}$, not 0.

Additional Practice

1. $xe^{x^2} = x\left(1 + x^2 + \dfrac{x^4}{2!} + \cdots + \dfrac{x^{2n}}{n!} + \cdots\right) = x + x^3 + \dfrac{x^5}{2!} + \cdots + \dfrac{x^{2n+1}}{n!} + \cdots$

2. (D)

3. (B) Antidifferentiation yields $g(x) = \left[7t - \dfrac{3}{2}(t-4)^2 + \dfrac{5}{3}(t-4)^3 + C\right)\Big]_4^x$, and evaluation at $x = 4$ gives $C = -28$.

Power Series

Content & Practice

1. $a = x; r = -3x; f(x) = \dfrac{x}{1+3x}$

2. (A) $2 - 3x + \dfrac{9}{2}x^2 - \dfrac{27}{4}x^3 + \cdots$

    (B) $4 + 4(3x - 1) + 4(3x - 1)^2 + 4(3x - 1)^3 + \cdots$

Additional Practice

1. (C) Note that $f(x) = \dfrac{x}{\sqrt{1-x^2}}$, the product of x and $g'(x)$. Therefore, a seventh-order

   power series of $f(x)$ can be found by $x\left(1 + \dfrac{x^2}{3} + \dfrac{3x^4}{8} + \dfrac{5x^6}{16}\right)$.

2. (D)

3. (A)

Radius and Interval of Convergence

Content & Practice

1. (A) $f(x) = \dfrac{1}{4-x}$

   (B) $2 < x < 4$

   (C) $x = 3$

   (D) Radius is 1.

2. (C)

3. $\dfrac{3}{2} < x < \dfrac{11}{2}$

4. $-4 < x < 6.$ The Ratio Test: $\displaystyle\lim_{n\to\infty}\left|\dfrac{\dfrac{(n+1)(x-1)^{n+1}}{5}}{\dfrac{n(x-1)^n}{5^n}}\right| < 1 \Rightarrow -1 < \dfrac{1}{5}(x-1) < 1$

Additional Practice

1. (A)

2. (C)

3. (E)

LaGrange Error Bound

Content & Practice

1. Error is less than or equal to $e^{0.5} \dfrac{(0.5)^3}{3!} \approx 0.0344$

2. $-1 + x + (x - 2)^2 + (x - 2)^3$. There is no error bound since there is no upper bound for the fourth derivative on the interval $|x - 2| < 1$.

Additional Practice

1. (E)

2. (B)

Practice Examinations
Calculus AB—Exam 1

Section I

Part A—No Calculator

Problem	Answer
1.	(B)
2.	(E)
3.	(D)
4.	(C)
5.	(D)
6.	(A)
7.	(B)
8.	(E)
9.	(C)
10.	(E)
11.	(A)
12.	(E)
13.	(B)
14.	(C)
15.	(A)
16.	(E)
17.	(C)
18.	(D)
19.	(D)
20.	(D)
21.	(B)
22.	(E)
23.	(A)
24.	(D)
25.	(D)
26.	(A)
27.	(E)
28.	(D)

Part B—Calculator Allowed

Problem	Answer
29.	(D)
30.	(B)
31.	(A)
32.	(D)
33.	(B)
34.	(E)
35.	(E)
36.	(C)
37.	(E)
38.	(D)
39.	(D)
40.	(A)
41.	(A)
42.	(B)
43.	(C)
44.	(D)
45.	(C)

Calculus AB—Exam 1: Section II, Part A

1. (A) $\dfrac{16}{3}$

    (B) $h \approx 1.520$

    (C) 8π

    (D) $k \approx 1.828$

2. (A) $\displaystyle\lim_{x \to -\infty} f(x) = 0$

    $\displaystyle\lim_{x \to \infty} f(x) = \infty$ or does not exist

    (B) Absolute minimum value of $f(x)$ is $f\left(-\dfrac{1}{3}\right) \approx -0.491$. $f'(x)4e^{3x}\left(3x + 1\right)0 \Leftrightarrow x = -\dfrac{1}{3}$.

    Since $f'(x) < 0$ for $x < -\dfrac{1}{3}$ and $f'(x) > 0$ for $x > -\dfrac{1}{3}$, $f(x)$ has a minimum at

    $x = -\dfrac{1}{3}$.

    (C) $(-0.491, \infty)$ Include left endpoint (but not exactly -0.491)

    (D) Absolute minimum value is $f\left(-\dfrac{1}{b}\right) = -\dfrac{a}{be}$.

3. (A)

    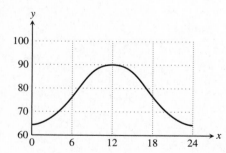

    (B) Ave. temp. $\approx 88°$F

    (C) $5.360 \leq t \leq 18.640$

    (D) Cost $\approx \$9.36$

Calculus AB—Exam 1: Section II, Part B

4. (A) Acceleration is positive on $(0, 45)$ and $(55, 60)$ because the velocity is increasing on $[0, 45]$ and $[55, 60]$.

   (B) Ave. accel. $\approx \dfrac{47}{60}$ ft/sec^2

   (C) $\dfrac{v(35) - v(25)}{10} = \dfrac{42 - 35}{10} = \dfrac{7}{10}$ ft/sec^2;

   $or \ \dfrac{v(35) - v(30)}{5} = \dfrac{42 - 40}{5} = \dfrac{2}{5}$ ft/sec^2;

   $or \ \dfrac{v(30) - v(25)}{5} = \dfrac{40 - 35}{5} = 1$ ft/sec^2

   (D) $\int_0^{60} v(t)\, dt \approx 2030$ ft. This is the total distance traveled in feet over the time 0 to 60 seconds.

5. (A) $\quad 2x + 6y\dfrac{dy}{dx} = 3x\dfrac{dy}{dx} + 3y$

   $6y\dfrac{dy}{dx} - 3x\dfrac{dy}{dx} = 3y - 2x$

   $\dfrac{dy}{dx}(6y - 3x) = 3y - 2x$

   $\dfrac{dy}{dx} = \dfrac{3y - 2x}{6y - 3x}$

   (B) $(h0) + (h1); y = \dfrac{2}{3}x - \dfrac{2}{3}, y = \dfrac{1}{3}x + \dfrac{2}{3}$

   (C) $(-2, -1)$ and $(2, 1)$

6. (A)

   (B) Slopes are positive for points where $x \neq 0$ and $y > -\dfrac{1}{2}$.

   (C) $y = \dfrac{1}{2}\left(11e^{x^3/3} - 1\right)$

Calculus AB—Exam 2

Section I

<table>
<tr><td colspan="2">Part A—No Calculator</td><td colspan="2">Part B—Calculator Allowed</td></tr>
<tr><td>Problem</td><td>Answer</td><td>Problem</td><td>Answer</td></tr>
<tr><td>1.</td><td>(C)</td><td>29.</td><td>(E)</td></tr>
<tr><td>2.</td><td>(A)</td><td>30.</td><td>(B)</td></tr>
<tr><td>3.</td><td>(C)</td><td>31.</td><td>(D)</td></tr>
<tr><td>4.</td><td>(C)</td><td>32.</td><td>(C)</td></tr>
<tr><td>5.</td><td>(E)</td><td>33.</td><td>(B)</td></tr>
<tr><td>6.</td><td>(C)</td><td>34.</td><td>(B)</td></tr>
<tr><td>7.</td><td>(E)</td><td>35.</td><td>(B)</td></tr>
<tr><td>8.</td><td>(B)</td><td>36.</td><td>(B)</td></tr>
<tr><td>9.</td><td>(C)</td><td>37.</td><td>(B)</td></tr>
<tr><td>10.</td><td>(C)</td><td>38.</td><td>(C)</td></tr>
<tr><td>11.</td><td>(D)</td><td>39.</td><td>(A)</td></tr>
<tr><td>12.</td><td>(B)</td><td>40.</td><td>(C)</td></tr>
<tr><td>13.</td><td>(A)</td><td>41.</td><td>(D)</td></tr>
<tr><td>14.</td><td>(C)</td><td>42.</td><td>(B)</td></tr>
<tr><td>15.</td><td>(B)</td><td>43.</td><td>(E)</td></tr>
<tr><td>16.</td><td>(C)</td><td>44.</td><td>(D)</td></tr>
<tr><td>17.</td><td>(A)</td><td>45.</td><td>(E)</td></tr>
<tr><td>18.</td><td>(B)</td><td></td><td></td></tr>
<tr><td>19.</td><td>(B)</td><td></td><td></td></tr>
<tr><td>20.</td><td>(E)</td><td></td><td></td></tr>
<tr><td>21.</td><td>(A)</td><td></td><td></td></tr>
<tr><td>22.</td><td>(D)</td><td></td><td></td></tr>
<tr><td>23.</td><td>(E)</td><td></td><td></td></tr>
<tr><td>24.</td><td>(B)</td><td></td><td></td></tr>
<tr><td>25.</td><td>(A)</td><td></td><td></td></tr>
<tr><td>26.</td><td>(D)</td><td></td><td></td></tr>
<tr><td>27.</td><td>(B)</td><td></td><td></td></tr>
<tr><td>28.</td><td>(E)</td><td></td><td></td></tr>
</table>

Calculus AB—Exam 2: Section II, Part A

1. (A) $x(t) = t^3 - 2t^2 - 4t + c$

   $x(1) = 1^3 - 2(1)^1 - 4(1) + c = 1 \Rightarrow c = 6$

   $x(t) = t^3 - 2t^2 - 4t + 6$

   (B) $\qquad v_{\text{avg}} = \dfrac{x(4) - x(0)}{4} = \dfrac{22 - 6}{4} = 4$

   $v(t) = 3t^2 - 4t - 4 = 4$

   $3t^2 - 4t - 8 = 0$

   $t = \dfrac{4 \pm \sqrt{(-4)^2 - 4(3)(-8)}}{2(3)} = \dfrac{4 \pm \sqrt{112}}{6} = \dfrac{4 + \sqrt{112}}{6} \approx 2.431$

   (C) $v(t) = 3t^2 - 4t - 4 = 0 \Rightarrow (3t + 2)(t - 2) = 0 \Leftrightarrow t = 2$

   Change of direction at $t = 2$ since v changes sign there.

   $x(0) = 6$ and $x(2) = -2$ and $x(4) = 22$

   Total distance $= |x(2) - x(0)| + |x(4) - x(2)| =$

   $\qquad\qquad\qquad |-2 - 6| + |22 - (-2)| = 8 + 24 = 32$

2. (A) Slope $= \dfrac{0 - 4}{\pi - \dfrac{\pi}{2}} = \dfrac{-4}{\dfrac{\pi}{2}} = \dfrac{-8}{\pi}$ \qquad Point $= (\pi, 0)$

   $y - 0 = \dfrac{-8}{\pi}(x - \pi)$ or $y = \dfrac{-8}{\pi}x + 8$

   (B) $f'(x) = 4 \cos x$

   Slope $= f'(\pi) = 4 \cos \pi = -4$ \qquad Point $= (\pi, 0)$

   $y = -4(x - \pi)$ or $y = -4x + 4\pi$

   (C) $4 \cos x = \dfrac{-8}{\pi} \Rightarrow x \approx 2.261$

   (D) Using the washer method: $\pi \int_{\pi/2}^{\pi} \left[(4 \sin x)^2 - \left(-\dfrac{8}{\pi}x + 8 \right)^2 \right] dx$

3. (A) $\dfrac{dv}{dt} = -3(v+5) \Rightarrow \dfrac{dv}{v+5} = -3\,dt \Rightarrow \int \dfrac{dv}{v+5} = \int -3\,dt$

$\ln|v+5| = -3t + c_1 \Rightarrow |v+5| = e^{-3t+c_1} \Rightarrow |v+5| = e^{-3t}e^{c_1}$

$$\Rightarrow v+5 = ce^{-3t} \Rightarrow v = ce^{-3t} - 5$$

$-15 = ce^{-3(0)} - 5 \Rightarrow -15 = c - 5 \Rightarrow c = -10 \Rightarrow v = -10e^{-3t} - 5$

(B) $\displaystyle\lim_{t\to\infty}(-10e^{-3t} - 5) = -10(0) - 5 \Rightarrow -5$ m/sec

(C) $-10e^{-3t} - 5 = -5.3 \Rightarrow -10e^{-3t} = -0.3 \Rightarrow e^{-3t} = 0.03 \Rightarrow t = \dfrac{\ln 0.03}{-3} \approx 1.169$

The crate will be able to land safely after about 1.169 seconds.

4.   (A)

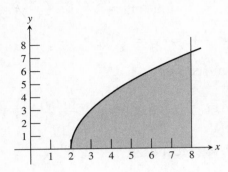

    (B) $\int_2^8 3(x - 2)^{1/2}\, dx = \left[3\left(\frac{2}{3}\right)(x - 2)^{3/2}\right]_2^8$

$$= \left[2(x - 2)^{3/2}\right]_2^8 = 2(6)^{3/2} - 2(0)^{3/2}$$

$$= 12\sqrt{6}$$

    (C) $A(k) = \int_2^k 3\sqrt{x - 2}\, dx$

    (D) $\dfrac{dA}{dk} = \dfrac{d}{dk}\left(\int_2^k 3\sqrt{x - 2}\, dx\right) = 3\sqrt{k - 2}$ (Fundamental Theorem of Calculus)

When $k = 8, \dfrac{dA}{dk} = 3\sqrt{8 - 2} = 3\sqrt{6}.$

5.   (A) $f'(x) = 3x^2 - 6x = 3x(x - 2)$

$f' \xleftarrow{\quad + + + \quad\big|\quad - - -\quad\big|\quad + + + \quad}$
$\phantom{f' \xleftarrow{\qquad\qquad}} 0 \qquad\quad 2$

There is a relative maximum at $x = 0$ because the function changes from increasing ($f'(x) > 0$) to decreasing ($f'(x) < 0$) there. There is a relative minimum at $x = 2$ because the function changes from decreasing ($f'(x) < 0$) to increasing ($f'(x) > 0$) there.

$f(0) = k$ is a relative maximum and $f(2) = k - 4$ is a relative minimum.

(B) In a cubic equation, we will only have three distinct roots if the relative maximum lies above the *x*-axis and the relative minimum lies below the *x*-axis (see the figure).

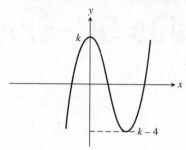

$$k > 0 \quad \text{and} \quad k - 4 < 0$$
$$0 < k < 4$$

(C) $f_{avg} = \dfrac{1}{1 - (-2)} \displaystyle\int_{-2}^{1} (x^3 - 3x^2 + k)\, dx = \dfrac{1}{3} \displaystyle\int_{-2}^{1} (x^3 - 3x^2 + k)\, dx$

$$= \frac{1}{3}\left[\frac{x^4}{4} - x^3 + kx \right]_{-2}^{1} = \frac{1}{3}\left[\left(\frac{1}{4} - 1 + k \right) - (4 + 8 - 2k) \right]$$

$$= \frac{1}{3}\left(3k - \frac{3}{4} - 12 \right) = k - 4.25$$

$$k - 4.25 = 2 \Longrightarrow k = 6.25$$

6. (A) $g(5) = \dfrac{1}{2}(3)(3) - \dfrac{1}{4}\pi(2)^2 = \dfrac{9}{2} - \pi$

(B) By the Fundamental Theorem of Calculus, we know that $g'(x) = f(x)$, the function whose graph is depicted. For the interval given, $g(x)$ has no endpoints and only one critical point, $x = 3$. Since $g'(x) = f(x)$ changes from positive to negative at that point, we know that $g(x)$ has a relative maximum at that point.

(C) $g(5) = \dfrac{9}{2} - \pi \Longrightarrow$ Point: $\left(5, \dfrac{9}{2} - \pi \right)$

From the graph, $f(5) = g'(5) = -2 \Longrightarrow$ Slope $= -2$

Tangent line: $y - \left(\dfrac{9}{2} - \pi \right) = -2(x - 5)$ or $y = -2x + \dfrac{29}{2} - \pi$

(D) Since g is differentiable for the interval $(-3, 7)$, g is continuous on $[-3, 7]$.

g has points of inflection at $x = 0$ and at $x = 5$ since g'' changes signs at both of these points.

Calculus BC–Exam 1

Section I

Part A – No calculator	
Problem	**Answer**
1.	(D)
2.	(A)
3.	(C)
4.	(B)
5.	(A)
6.	(E)
7.	(E)
8.	(D)
9.	(B)
10.	(A)
11.	(B)
12.	(A)
13.	(D)
14.	(E)
15.	(A)
16.	(D)
17.	(E)
18.	(B)
19.	(A)
20.	(B)
21.	(E)
22.	(C)
23.	(D)
24.	(A)
25.	(B)
26.	(A)
27.	(E)
28.	(C)

Part B – Calculator allowed	
Problem	**Answer**
29.	(B)
30.	(A)
31.	(E)
32.	(D)
33.	(B)
34.	(C)
35.	(B)
36.	(A)
37.	(D)
38.	(D)
39.	(A)
40.	(C)
41.	(C)
42.	(B)
43.	(E)
44.	(E)
45.	(D)

1. (A) $v(0) = \langle 2, 5 \rangle$; Speed $= |v| = \sqrt{2^2 + 5^2} = \sqrt{29}$; $a = \left\langle \dfrac{d^2x}{dt^2}, \dfrac{d^2y}{dt^2} \right\rangle$; By nDeriv,

$a = \langle 0, 1 \rangle$.

(B) Since $V(0) = \langle 2, 5 \rangle$, $\dfrac{dy}{dx} = \dfrac{5}{2}$ $y - 4 = \dfrac{2}{5}(x - 3)$

(C) $D = \int_0^2 \sqrt{(t^2 + 4) + (3e^t + 2e^{-t})^2}\, dt = 21.455$

(D) $x = 3 + \int_0^2 \sqrt{t^2 + 4}\, dt = 7.591$

2. (A) $f(3) = 8$;

$$\frac{f^{\mathrm{IV}}(3)}{4!} = 9;$$

$$f^{\mathrm{IV}}(3) = 9 \cdot 4! = 216$$

(B) Differentiate the first five terms of $P(x)$, term by term.
$$P'(x) = -4 + 10(x - 3) - 21(x - 3)^2 + 36(x - 3)^3;$$
$$f'(3.2) \approx -4 + 10(0.2) - 21(0.2)^2 + 36(0.2)^3 = -2.552$$

(C) $P_6(h, x) = \int_3^x P_5(t)\, dt$

$= \int_3^x (8 - 4(t - 3) + 5(t - 3)^2 - 7(t - 3)^3 + 9(t - 3)^4 - 6(t - 3)^5)\, dt$

$= 8(x - 3) - 2(x - 3)^2 + \dfrac{5}{3}(x - 3)^3 - \dfrac{7}{4}(x - 3)^4 + \dfrac{9}{5}(x - 3)^5 - (x - 3)^6$

(D) No. We only have information about $f(3)$ and the first five derivatives of f evaluated at 3. We can only estimate $f(4)$.

3. (A) Find the positive point of intersection, $x = 2.0423585$ and store to A. Symmetry makes the other point of intersection $-A$.

$$\int_{-A}^{A} \left(4 - \frac{1}{2}x^2 \right) - \sec\frac{x}{2}\, dx = 8.434 \text{ or } 8.435$$

(B) Using the washer method, the outer radius is $y = 4 - \frac{1}{2}x^2$.

$$V = \pi \int_{-A}^{A} \left[\left(4 - \frac{1}{2}x^2 \right)^2 - \left(\sec \frac{x}{2} \right)^2 \right] dx$$

$$= 39.661\pi \text{ or } 39.662\pi$$

$$\approx 124.600$$

(C) Use symmetry to double the first quadrant length. Also use the formula

$$L = \int_a^b \sqrt{1 + \left(\frac{dy}{dx} \right)^2} \, dx.$$

$$\frac{d}{dx} \left(4 - \frac{1}{2}x^2 \right) = -x; \quad \frac{d}{dx} \left(\sec \frac{x}{2} \right) = \frac{1}{2} \sec \frac{x}{2} \tan \frac{x}{2};$$

$$L = 2\int_0^A \sqrt{1 + (-x)^2} \, dx + 2\int_0^A \sqrt{1 + \left(\frac{1}{2} \sec \frac{x}{2} \tan \frac{x}{2} \right)^2} \, dx$$

Calculus BC—Exam 1: Section II, Part B

4. (A) $\dfrac{d}{dx}(4x - 2x^2y) = \dfrac{d}{dx}(y^2 + 1)$

$$4 - 2\left(x^2\dfrac{dy}{dx} + 2xy\right) = 2y\dfrac{dy}{dx} + 0$$

$$4 - 2x^2\dfrac{dy}{dx} - 4xy = 2y\dfrac{dy}{dx}$$

$$2 - x^2\dfrac{dy}{dx} - 2xy = y\dfrac{dy}{dx}$$

$$2 - 2xy = y\dfrac{dy}{dx} + x^2\dfrac{dy}{dx}$$

$$2 - 2xy = (y + x^2)\dfrac{dy}{dx} \Longrightarrow \dfrac{dy}{dx} = \dfrac{2 - 2xy}{y + x^2}$$

(B) $4x - 2x^2y = y^2 + 1$ with $x = 1$ becomes $4 - 2y = y^2 + 1$.

$0 = y^2 + 2y - 3 \Leftrightarrow y = -3$ or $y = 1$. At $(1, 1)$, $\dfrac{dy}{dx} = \dfrac{2 - 2 \cdot 1 \cdot 1}{1 + 1^2} = 0$.

(C) $\dfrac{d}{dx}\left(\dfrac{2 - 2xy}{y + x^2}\right) = \dfrac{(y + x^2)\left[0 - 2\left(x\dfrac{dy}{dx} + y\right)\right] - (2 - 2xy)\left(2x + \dfrac{dy}{dx}\right)}{(y + x)^2}$

At $(1, 1)$, $\dfrac{dy}{dx} = 0$ and $2 - 2xy = 0$.

$$\dfrac{d^2y}{dx^2} = \dfrac{(y + x^2)(0 - 2y) - 0}{(y + x^2)^2} = \dfrac{2(-2)}{2^2} = -1$$

$\dfrac{dy}{dx} = 0$ and $\dfrac{d^2y}{dx^2} < 0$, by the second derivative test for extrema there is a local maximum at $(1, 1)$.

5. (A) $h(4) = \int_0^4 g(t)\, dt$

$$\dfrac{1}{4}\pi \cdot 3^2 + \dfrac{1}{2} \cdot 1 \cdot -2 = \dfrac{9}{4}\pi - 1$$

(B) Identify where $h'(x)$ changes from negative to zero to positive, and check endpoints of the domain. By the Fundamental Theorem of Calculus, $h'(x) = \frac{d}{dx}\int_0^x g(t)\,dt = g(x)$.

At $x = 5$, $g(x) = h'(x) = 0$. Since $h'(x)$ changes from negative to zero to positive at $x = 5$, a local minimum of g exists at $x = 5$.

$h(-2) = \int_0^{-2} g(t)\,dt = -3$; $h'(x) = g(x) > 0$ for $-2 < x < 3$ so h is increasing on that interval. Therefore, $x = -2$ is also a local minimum.

(C) $h'(2) = g(2)$; for $0 \le x \le 3$, $g(x) = \sqrt{9 - x^2}$. Thus $h'(2) = g(2) = \sqrt{5}$.

(D) Find the x-value of each local extremum for $h' = g$ on the interval. By observation, these are $x = 0$ and $x = 4$ ($-2 + 6$ are excluded). Thus the graph of h has inflection points where $x = 0$ and where $x = 4$.

6. (A)

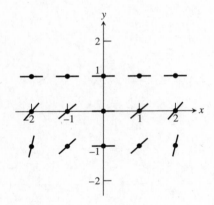

(B) $$\frac{dy}{dx} = x^2(1 - y)^2$$

$$\frac{dy}{(y - 1)^2} = x^2\,dx$$

$$-\frac{1}{y - 1} = \frac{1}{3}x^2 + C, \text{ at } (3, 0), 1 = \frac{1}{3}\cdot 3^3 + C, \text{ so } C = -8.$$

Thus $y = \dfrac{1}{8 - \frac{1}{3}x^3} + 1 \left(\text{or } y = \dfrac{9 - \frac{1}{3}x^3}{8 - \frac{1}{3}x^3} \right)$

(C) $\lim\limits_{x \to \infty} f(x) = \lim\limits_{x \to \infty}\left(\dfrac{1}{8 - \frac{1}{3}x^3} + 1 \right) = \dfrac{1}{\infty} + 1 = 1.$

For $y = 1$, the slope is always 0.

The line $y = 1$ is an asymptote of the graph of f.

Calculus BC—Exam 2

Section I

	Part A—No Calculator			Part B—Calculator Allowed	
Problem	**Answer**			**Problem**	**Answer**
1.	(A)			29.	(E)
2.	(C)			30.	(B)
3.	(C)			31.	(C)
4.	(E)			32.	(A)
5.	(B)			33.	(C)
6.	(D)			34.	(D)
7.	(A)			35.	(D)
8.	(E)			36.	(D)
9.	(D)			37.	(A)
10.	(C)			38.	(E)
11.	(A)			39.	(C)
12.	(E)			40.	(B)
13.	(D)			41.	(A)
14.	(B)			42.	(A)
15.	(A)			43.	(B)
16.	(C)			44.	(E)
17.	(B)			45.	(D)
18.	(B)				
19.	(E)				
20.	(D)				
21.	(A)				
22.	(C)				
23.	(E)				
24.	(A)				
25.	(C)				
26.	(D)				
27.	(C)				
28.	(C)				

Calculus BC—Exam 2: Section II, Part A

1. (A)

    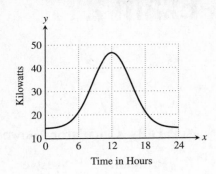

    Time in Hours

    (B) $\dfrac{1}{18-6}\displaystyle\int_6^{18} K(t)\, dt = 37.732$

    (C) $25 - 20 \cos \dfrac{\pi t}{12} > 35$ for $8 < t < 16$

    (D) $0.07 \displaystyle\int_8^{16} 25 - 20 \cos \dfrac{\pi t}{12}\, dt = \$23.26$

2. (A) Use $dy/dx = x^2 y$ to calculate the slope at each of the given points and then sketch lines with these slopes.

    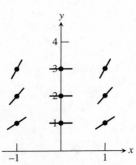

    (B)

x	y	$f'(x)$
0	1	0
0.1	1	0.01
0.2	1.001	

(C) $\dfrac{dy}{dx} = x^2 y$

$\dfrac{dy}{y} = x^2$

$\ln |y| = \dfrac{x^3}{3} + C$

$y = Ce^{x^3/3}$

$1 = Ce^0$

$C = 1$

$y = e^{x^3/3}$

$e^{(0.2)^3/3} = 1.003$

3.   (A) $7 - 4x + \dfrac{1}{2}x^2 + x^3; f(0.3) = 5.872$

(B) Substitute x^2 for x in the first three terms of the expression $7 - 4x + \dfrac{1}{2}x^2 + x^3$ to obtain $7 - 4x^2 + \dfrac{1}{2}x^4$.

(C) $\displaystyle\int_0^x \left(7 - 4t + \dfrac{1}{2}t^2 + t^3 \right) dt = 7x - 2x^2 + \dfrac{x^3}{6} + \dfrac{x^4}{4}$

(D) $h(1)$ cannot be determined because $f(t)$ is known only for $t = 0$ and $t = 1$.

4. (A) $\int_0^2 (4 - x^2)\, dx = \dfrac{16}{3}$

    (B) Use the method of disks about the x-axis: $\int_0^2 \pi(4 - x^2)^2\, dx = \dfrac{256}{15}\, \pi.$

    (C) Use the method of disks about the y-axis: $\int_0^4 \pi(4 - y)\, dy = 8\pi.$

5. (A) Domain: $(0, \infty)$

    (B) $f'(x) = 2(1 + \ln x)$. Setting the derivative equal to 0 gives $x = \dfrac{1}{2e}$ and $y = -\dfrac{1}{e}$.

    Use the second derivative test: $f''(x) = \dfrac{2}{x}$ so $f''\left(\dfrac{1}{2e}\right) = 4e > 0$. Since the second deriva-

    tive is positive, there is a minimum at $x = \dfrac{1}{2e}$. Since this is the only critical point in the

    domain, $y = -\dfrac{1}{e}$ is an absolute minimum.

    (C) $\displaystyle\lim_{x \to \infty} 2x \ln 2x = \infty;$ Range: $\left[-\dfrac{1}{e}, \infty\right)$

    (D) $f'(x) = b(1 + \ln bx)$; Setting the derivative equal to zero gives $x = \dfrac{1}{be}$ and $y = -\dfrac{1}{e}$.

    Use the second derivative test: $f''(x) = \dfrac{b}{x}$ so $f''\left(\dfrac{1}{be}\right) = b^2 e > 0$. Since the second deriv-

    ative is positive, there is a minimum at $x = \dfrac{1}{2e}$. Since this is the only critical point in the

    domain, $y = -\dfrac{1}{e}$ is an absolute minimum.

6. (A) $\dfrac{dx}{dt} = (t + 1)^{-1/2}$

$x = 2(t + 1)^{1/2} + C$

Initial conditions:

$-1 = 2\sqrt{4} + C$

$C = -5$

$x = 2\sqrt{t + 1} - 5$

(B) $\dfrac{dy}{dt} = 2x\dfrac{dx}{dt} - 2\dfrac{dx}{dt}$

$\dfrac{dy}{dt} = \dfrac{2\left(2\sqrt{t + 1} - 5\right)}{\sqrt{t + 1}} - \dfrac{2}{\sqrt{t + 1}}$

$\dfrac{dy}{dt} = \dfrac{4\sqrt{t + 1} - 12}{\sqrt{t + 1}}$

(C) Location: $(1, -1)$

$\dfrac{dx}{dt} = \dfrac{1}{3}$

$\dfrac{dy}{dt} = 0$

Speed $= \sqrt{\left(\dfrac{1}{3}\right)^2 + 0^2} = \dfrac{1}{3}$